THE SILVERY TAY

Keith Brockie

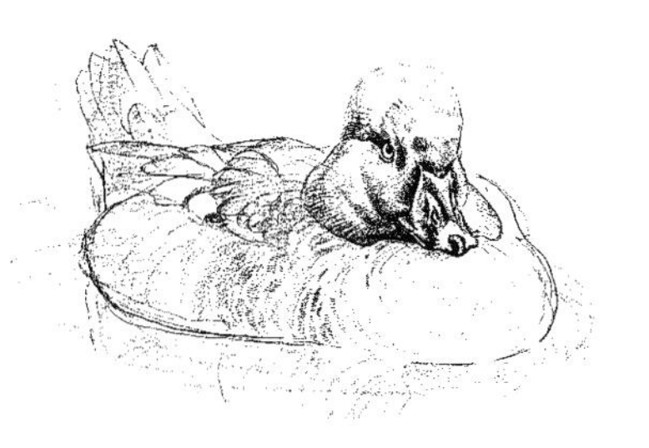

ALSO BY KEITH BROCKIE

Keith Brockie's Wildlife Sketchbook, Dent 1981
One Man's Island, Dent 1984

Illustrated by Keith Brockie

The Puffin (M.P. Harris), Calton 1984
The Sparrowhawk (I. Newton), Calton 1986
Studies on the Tihamah (edited by Francine Stone), Longman 1985

KEITH BROCKIE

THE SILVERY TAY

Paintings and Sketches from a Scottish river

J.M. DENT AND SONS LTD
London

First published in Great Britain 1988

Designed by Gaye Allen

This book is phototypeset in 12/14 Apollo
by Keyspools Ltd, Golborne, Lancs
Colour separations by Newsele Litho Ltd
Printed in Italy by Printers Srl, Trento
Bound by L.E.G.O., Vicenza
for
J.M. Dent & Sons Ltd,
91 Clapham High Street, London SW4 7TA

British Library Cataloguing in Publication Data

Brockie, Keith, *1955–*
The silvery Tay,
1. Scotland. Tay River. Organisms
I. Title
574.9412′8

ISBN 0–460–04726–4

Contents

Headwaters

Lochs

48 Mallards at sunset R, 2, P, WC
49 Mallards on ice V, 1, P, WC
50 Green-winged Teal V, 1, P, WC
51 Goldeneyes displaying V, 2, P, WC
52 Goldeneye female F, 1, P
53/54 Goldeneye 762 × 508 mm S, 2, P, WC
55 Snipe V, 1, P, WC
56 Lapwing V, 1, P, WC
57 Spotted Redshank R, 2, P, WC
58 Pink-footed Goose × 1.5 F, 2, P, WC
59/60 Cormorants V, 2, P, WC
61/62 Dippers F, 2, P, WC

Main Rivers

63 Small White Orchid F, 1, P, WC
64 Greater Butterfly Orchid F, 1, P, WC
65 Fragrant Orchid F, 1, P, WC
66 Spotted Orchid F, 1, P, WC
67 Moths S, 1, P, WC
68 Long-eared Bat F, 2, P, WC
69 Brown Hare V, 1, P
70 Brown Hare × 1.5 S, 2, P, WC
71 Roe Buck sketches F, 1, P
72 Roe Buck 762 × 508 mm S, 2, O
73 Salmon sketches F, 1, P, WC
74 Salmon leaping S, 1, P, WC
75 Salmon female F, 1, P, WC
76 Salmon male S, 1, P, WC
77/78 Salmon leaping the falls 914 × 609 mm S, 2, O
79 Salmon underwater F, 1, P, WC
80 Greater Black-backed Gull, dead Salmon V, 1, P, WC
81/82 Sea Trout A, 1, P, WC
83 Grayling A, 1, P, WC
84 Mallard F, 1, WC
85 Heron H, 1, P
86 Heron H, 1, P, WC
87 Heron adult half asleep H, 1, P, WC
88 Heron portrait H, 1, P, WC
89 Heron H, 1, P, WC
90 Heron H, 1, P
91/92/93 Heron chicks H, 1, P, CP, WC
94 Heron H, 1, P, CP
95 Goosander female on nest F, 1, P, WC
96 Goosander nest site F, 1, P, WC
97 Goosanders, Red-breasted Merganser V, 1, P, W
98 Goosander in rapids V, 1, P, WC
99 Common Tern H, 1, P, WC
100 Common Gull H, 1, P, CP, WC
101 Lapwing V, 1, P, WC
102 Common Sandpiper H, 1, P, WC
103 Ringed Plover on nest H, 1, P, WC
104 Ringed Plover chicks F, 1, P, CP, WC
105 Oystercatcher portrait H, 1, P, WC
106 Oystercatcher with chick V, 1, P, WC
107 Sand Martin colony F, 1, P, WC
108 Sand Martin nest hole F, 2, P, WC
109 Grey Wagtail S, 2, P, WC
110 Swallows collecting mud S, 1, P, WC
111/112 Swallows, Martins on roof V, 2, P, WC
113 Cormorant, Goldeneye V, 1, P, WC
114 Whooper Swan sketches F, 1, P
115/116 Whooper Swan family × 1.5 S, 2, P, WC

Estuary

117 Grayling, Thyme — F, 1, P, WC
118 Dark Green, Fritillary, Willowherb — F, 1, P, WC
119 Centaury, Burnet Moths — F, 1, P, WC
120 Grass of Parnassus — F, 1, WC
121 Coralroot Orchid — F, 1, P, WC
122 Creeping Lady's Tresses — F, 1, P, WC
123 Lesser Twayblade — F, 1, P, WC
124 Common Twayblade — F, 1, P, WC
125 Yellow Birdsnest — F, 1, P, WC
126 Moonwort, Adders tongue — F, 1, P, WC
127/128 Brent Geese — V, 1, P, WC
129 Curlews bathing — H, 1, P, WC
130 Oystercatchers at roost — H, 1, P, WC
131 Common Scoter — H, 2, P, WC
132 Heron — H, 1, P, CP, WC
133 King Eider male — H, 2, P, WC
134 Common Eider pair — H, 1, P, WC
135 Red-breasted Merganser — H, 1, P
136 Mute Swan — S, 1, P, WC
137 Cormorants — H, 2, P, WC
138 Cormorant drying out — H, 2, P, WC
139 Cormorant adult — H, 1, P, WC
140 Cormorant immature with fish — H, 1, P, WC
141 Little, Gulls — F, 2, P, WC
142 Arctic Tern — F, 2, P, WC
143 Bar-tailed Godwits — H, 2, P, WC
144 Grey Plovers — H, 2, P, WC
145/146 Sandwich Terns × 1.5 — F, 2, P, WC
147 Greater Black-backed Gull — H, 2, P, CP, WC
148 Ringed Plover — V, 1, P, WC
149 Common Seals — F, 2, CP
150 Common Seal Pup — F, 2, P, CP, WC

LEGEND

1 Primary (original) drawing/painting
2 Secondary work taken from a primary drawing
P Pencil
CP Coloured Pencil
WC Water Colour (vis White Body Colour)
O Oil

Mode of Work

A = Aquarium, fish tank
F = On Foot
H = Portable canvas hide
R = Reserve observation hide
S = Studio
V = Using vehicle as a hide

All work reproduced same-size unless specified otherwise.

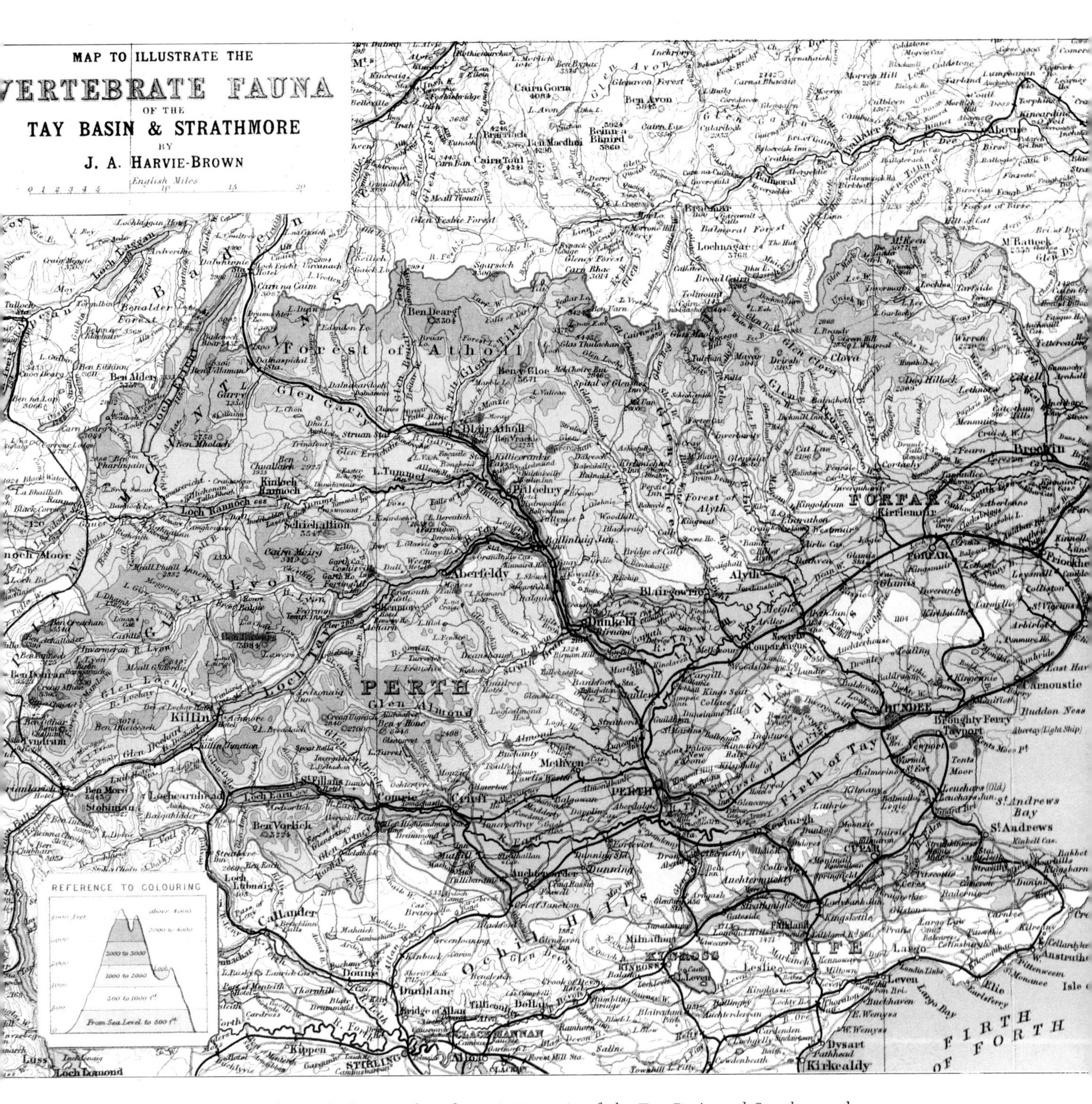

Above, an old colour relief map taken from *A Fountain of the Tay Basin and Strathmore*, by J. A. Harvie-Brown, published in 1906. Although much of the human infrastructure has changed, the map illustrates better than more modern maps the form of the Tay basin landmass. Except for some sensitive locations I have marked the areas where I have worked with red dots.

Opposite, a satellite map showing Scotland (except Shetland) with parts of Northern Ireland and England.

INTRODUCTION

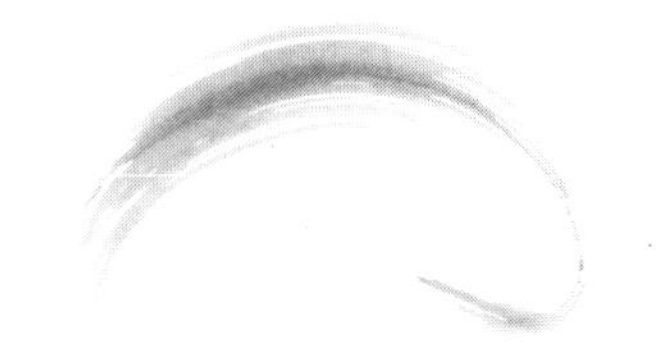

The Tay watershed is the largest catchment area in Scotland, encompasing some 2,800 square miles. The river and its surrounds form the silvery thread – a thread of life – which ties this book together. I have organised the drawings into four main habitat types. First the headwater areas based mainly on Ben Lawers, the highest mountain in the Tay basin and famed for its rich alpine flora. Second the wildlife of the many lochs, of varying size and character, within the region. The third section features the river Tay and the fauna found in and around its huge network of tributaries. Last but not least comes the teeming birdlife, seals and flora of the Tay estuary and Tentsmuir. With such a vast area to work in I selected a representative sample of those subjects which interested me most from an artistic and educational point of view. The book is therefore not a species-by-species account. Nor, I regret, did I have enough time to paint much of the landscape; in any case, the Tayside, which is one of the most beautiful areas in Scotland, deserves a book on its landscapes alone.

The actual source of the Tay is on the north-east slope of Ben Lui (Beinn Laoigh), a fine mountain some 1,130 metres (3,708 ft) high. Various streams merge to form the River Cononish flanked by the hills of Beinn Dubhchraig and Beinn Chuirn. Near Tyndrum this becomes the River Fillan which flows into Loch Dochart and Loch Iubhair before emerging as the River Dochart. This flows east to join Loch Tay at Killin. From this huge loch the River Tay then emerges at Kenmore flowing past the grandeur of Taymouth Castle down the fertile valley of Strath Tay flanked by some fine old deciduous woodland. Some fifteen miles below Loch Tay, by Ballinluig, the Tay is joined by the River Tummel, and doubles in size. From here the Tay flows some 31 miles to Perth via Dunkeld through magnificent policy woodland and rich farmland. Along its course it is further swollen by the Rivers Braan, Isla and Almond. At Perth the river becomes tidal and, after another 31 miles, the Firth of Tay reaches the open sea past Dundee. In all it is 117 miles long, and has a larger flow of water than any other river in Britain.

Most of the large lochs and their rivers have been harnessed in some way for hydro-electric power. Some of these schemes, involving tunnels, stream diversions, dams and power stations, have prevented salmon reaching the upper river spawning grounds. In other schemes, fish passes have been constructed to allow fish to surmount obstacles such as dams. The most famous fish ladder is at

Pitlochry Dam which created Loch Faskally. A great tourist attraction, this fish pass is nearly 250 metres long with some 35 pools. An underground chamber with a viewing window allows people to watch ascending salmon swimming through one of the pools when they are running. An electronic recorder counts the fish as they pass upstream and the yearly average during the 1970s was 7,714 salmon and grilse.

The Tay river system has long been famous for its salmon fishing. Indeed the British record for a rod-caught salmon was set on the Tay by Miss Georgina Ballantine back in October 1922. This monster cock fish weighed in at 64 lbs. The 1986 Tay District Salmon Fishers Board report gives that season's catch on the Tay system as 57,311 fish: rods accounted for some 10,323 fish, the remainder being taken by commercial nets. The season for rod-caught fish runs from 15 January to 15 October. Salmon spawn in gravel beds known as redds in the late autumn and early winter. They hatch in the early spring as alevins, becoming known as parr after their first year. In their second to fourth year they turn silver, are called smolts, and then migrate to the sea. Fish tagged as smolts in Scottish rivers have been recovered off Greenland. Most of the salmon return to their natal river after two to four years at sea and ascend at varying speeds according to the time of year and river condition. The grilse return after little more than a year at sea. The salmon seldom feed upon re-entering fresh water and their silvery appearance darkens considerably as spawning time approaches. At this time the male's lower jaw develops a hook known as a kype. After spawning, the fish are known as kelts and many die at this stage, although some females do return to spawn a second time.

The salmon – this king of all fish – engenders fierce loyalties from concerned bodies. Controversy continues over the commercial netting of salmon in the tidal waters, which employs relatively few people. In comparison the rod-caught fish bring in a much larger income to the local human infrastructure and tourist economy. Attempts are being made to buy out the netting stations and reduce the netting hours to allow more fish to enter the river system. Natural predators are often blamed for the reduction in salmon numbers, but this is much too simple. It overlooks man's greed – such depredations as the netting of their Greenland feeding waters, illegal offshore drift netting, river pollution and large-scale poaching. The grey and common seal, cormorant, goosander, red-breasted merganser, heron and osprey are all, to some extent, persecuted for the crime of taking salmon. Rogue seals do occur at fixed coastal nets and specific licences are issued for the shooting of them and some species of birds but much illegal shooting still goes on. Let us not forget that the aforementioned species of birds also take fish such as pike, perch, trout and eels which eat young salmon. At one fish farm near Comrie the owner was recently prosecuted for poisoning herons with fish doped with alphachlorolose. In a pile of burnt remains, including a Danish ringed bird and one I had ringed, some eight heron rings were found. This tally indicated that a much higher number of birds had been killed, because only a small percentage of herons are ringed. It isn't necessary to resort to such drastic measures. Better design of fish-rearing ponds can eliminate a great deal of heron predation.

Although I can see part of the Tay estuary from my house I have more tangible links with the river. A tributary of the Tay flows through my garden in front of the house. Here I have constructed a concrete dam to form a pond for my captive wildfowl. At present this collection includes a pair of whooper swans, forty ducks of various kinds, geese and a common gull. I keep the wildfowl as models for my drawing and painting as well as for the constant pleasure of watching and listening to them. Indeed at times they can be quite a distraction from my studio window as I attempt to concentrate on some more mundane work. They all have their different characters and idiosyncrasies. Also, in artistic terms, I can learn much from the play of light on their various plumages.

The pond, stream and marshy ground attract lots of other wildlife as well. Regulars include the lively pied and grey wagtails, waterhens and mallard. Talking of mallard, the female which I sketched on her nest near the bottom of my garden met a most unfortunate end. Whilst off feeding she made the fatal mistake of landing within my golden eagle's pen (he is disabled) and was duly pounced on! Summer comes alive as swallows and martins hawk for flies above the pond and gather mud and feathers for their nests. The swallows collect the wildfowl's moulted feathers and time and time again they pluck the feathers from the ground and fly up into the air, only to drop them, swoop down and catch them again before they reach the ground. Whoever said that birds never have fun!

During the harsher winter months many finches and buntings are drawn to the easy pickings in the wildfowl feeding-trough. In turn this attracts daily assaults by the sparrowhawks. Recently, whilst I was looking out of my window in search of inspiration, a lovely male merlin caught a sparrow in the garden right in front of me. Other occasional visitors to my pond include species such as heron, snipe and dipper. Mammal species include the pipistrelle bats which hunt for flies over the pond during the twilight hours. Stoats often search the stream margins for rats attracted by the wildfowl food. Fortunately I haven't yet been visited by its mustelid cousin, the mink, which could cause serious damage to my wildfowl. Mink are now well established throughout the Tay river system.

I enjoy tremendously the diverse wildlife of Tayside through watching and painting it, and I hope I can impart some of this pleasure to the readers of this book. I can't bring alive the sounds, smells and broader environment of the Tay, just try to give a little insight into a transitory moment in time for its flora and fauna. Sketching and painting wild birds and animals must be one of the most difficult of artistic disciplines, for seldom can one arrange to find a long-term stationary model, in the pose one wants. It helps, however, at least to know how, when and where to find one's subjects. With much of my work for this book I travelled to specific locations to look for subjects. Otherwise I ventured out on the off-chance that I might spot something interesting.

Time and the weather largely dictated my movements and what I could draw, and I had to choose those subjects which interested me but didn't impose too many insuperable difficulties. For example, sketching otters in the wild would have taken up too many twilight hours to have achieved anything meaningful. As it was, I didn't see any wild otters during my extensive fieldwork, although I kept finding their spraints (droppings) in various riverside locations. There were other disappointments, too. I missed some rare and interesting species within the area – 'dipping out' as the 'twitching' terminology goes. An immature sea eagle was present in an upland area for a few weeks but, although I searched far and wide, I couldn't locate it. All I found were moulted feathers and pellets (remains of mountain hares) where it had been roosting. This was a wing-tagged bird, one of the sea eagles brought in from Norway as part of a reintroduction scheme by the Nature Conservancy Council. Initially released on the island of Rhum these magnificent eagles are now breeding successfully on the west coast of Scotland. Another bird I failed to see was a red-rumped swallow, a rare migrant more usually found in the Mediterranean. This bird had been frequenting a street bordering the beach at Broughty Ferry between 7 and 8 November, but, as luck would have it, it departed some twenty minutes before I arrived!

I hope people can see an evolution in my work throughout the three books I have produced. I'm always happy with what I have done at the time, but never wholly satisfied. If I was, it would be the moment to give up, for it would mean the end of the addictive agony of striving for some sort of aesthetic perfection. My joy in nature is an intensely private and emotional one, but I always hope that what I do captures on paper some of my innermost feelings. Much of my time in the field is spent in contemplative watching – waiting for a flash of inspiration for some composition or a particular pose. Sketching can be a frenetic, nervous experience especially if one is trying to capture some fleeting moment. Art to me heightens and deepens human awareness and brings another dimension to the natural world. I'm not consciously trying to do something different. I simply don't pretend to understand some of the modern artwork whose hidden meanings critics often muse over. Sketching and painting amongst wildlife is what I revel in, hopefully evolving through greater familiarity with technique and subject.

If this book helps more people to appreciate their natural heritage then it will have achieved one of its aims. As man's greed and technological advances ever more rapidly impoverish our natural environment, a greater conservation lobby is necessary to ensure its survival for future generations. All I can hope is that I may have helped in this process.

Acknowledgements

I owe a debt of gratitude to the many people who have helped me in the completion of this book, especially to Steve Cooper and Allan Allison for their help and companionship in the field. Also special thanks to Tom Adamson, fishing ghillie at Kinkell; Anita Burch, summer warden for the Scottish Wildlife Trust at Keltneyburn; Euan Cameron; Claire Geddes; Pete Kinnear, Nature Conservancy Council warden for Tentsmuir N.N.R.; Dave Mardon, National Trust for Scotland warden for the Lawers area; A. Steedman; Forestry Commission; Tillhill Forestry and other landowners; and the Telegraph Colour Library for permission to reproduce the Landsat map. Peter Shellard of J.M. Dent provided much help in the planning stages and enthusiasm throughout. Pete Moore took the photographs except for the osprey, eagle and Lucky Scalp shots. Steve Cooper took the one of myself at an osprey nest and Stuart Rae photographed the eagle shots. Last but not least, thanks to my wife Morag for bearing with my unorthodox hours and temperamental moods during inclement weather – the latter all too frequent during the 1987 'summer'.

Protected Species

Some of the species, namely the osprey, black-throated diver and long-eared bat are specially protected by law and should on no account be disturbed (birds during the breeding season) without an appropriate permit from the National Environment Research Council. In addition raptors such as the golden eagle, peregrine falcon and merlin (injured or otherwise) should not be kept by anyone except when licensed to do so by the Department of the Environment.

WORKING METHODS

Few sketching positions are as awkward as the one above, in this case an osprey eyrie near a well-grown chick (though I remember equally well the discomfort of sitting in a climbing harness atop a tall Scots pine in Finland sketching a confiding female pine grosbeak on its nest!). I also recall how once, in a heron hide, after many hours in a cramped seated postion, I tried to stand up but my legs gave way. I toppled back and nearly fell right off the platform to the ground below. Fortunately I managed to grab one of the hide poles in time. It was a salutary reminder to get the blood flowing after sketching from a confined position, before moving around too much!

For any painter of wild birds and animals the most important piece of equipment is an extension of one's eyes – preferably a telescope. It is invaluable for the close observation of normally distant species. I use one with a close-focusing facility that enables me to do intimate work from the hide. Binoculars are passable, but much less effective. A telescope on a tripod or car window mount leaves one's hands free to draw without the hassle of constantly manoeuvering binoculars. A 30 × magnification and a good-sized object lens are perfectly adequate for most situations. With larger magnifications one tends to lose the quality of light.

Hides are generally excellent places to work from. Despite its bright colour my Volkswagen camper makes a good hide because it is some way above the ground. In bad weather it is also a spacious vehicle in which to work. Permanent reserve hides are good venues especially if you can use them at odd hours when few people are about; occasionally the subjects I've been drawing have fled because of the unthinking, noisy behaviour of people coming into hides. A portable canvas hide is an excellent way of getting close to such wildlife locations as nesting birds or a roost site. So long as I take great care, I can get very near to my subjects without disturbing them. My portable hide is just big enough to take me, a telescope and tripod, a collapsible seat and sketchbooks or painting blocks. Sufficient light can be introduced through side windows covered by flaps. At the moment I am modifying the roof with a clear material so that I can paint more easily in dull and overcast conditions.

Plants offer less of an artistic challenge than birds or animals simply because they don't move. The only problem is finding them, particularly the rarer species – and of course waiting for a suitable 'window' in the weather in which to paint them. In this book I have tried to show plants in their habitat as far as possible, using the backdrop to highlight the flowers. When I'm engaged on botantical work insects are usually the worst nuisance to deal with, often driving me to distraction; stationary for hours, I am truly a sitting target. Tentsmuir was especially bad in this respect, with hordes of biting flies such as clegs and mosquitoes. To counteract them I covered myself with the full-length mosquito net which I had last used on an expedition to the North Yemen. Using my tripod as an overhead anchor point I pegged the sides to the ground surrounding the plant. Thus I had a relatively trouble-free environment to paint in – except that I then had to cope with the crawling ants!

Some of the fish, such as the sea trout, I placed in a glass tank with an aeration pump. Whenever a friend caught a fish for me I put it straight into a bucket and rushed back to the fishing hut to drop it in the tank. They were ideal models because, after some initial thrashing about, they tended not to move much. Compared with a specimen fresh out of the water they look much flatter underwater, with few highlights. Dead fish discolour very quickly so, in such cases, one has to paint them almost immediately.

Basic field sketching/painting equipment

Container with a selection of pencils, 6B to H; pair of dividers for measuring specimens; scalpel for sharpening pencils etc; rubber; selection of sable watercolour brushes in a plastic holding tube; watercolour sketching box with a selection of half-pan artist's watercolours; tube of permanent white gouache; metal box with a range of coloured pencils, water soluble; plastic water container; old army surplus gas mask bag with lots of useful compartments; waterproof canvas holder for watercolour blocks; sketchbook for quick drawings of subjects and initial composition ideas; choice of pre-stretched watercolour blocks with different grains of paper; paper stretched on a board with gumstrip.

My vehicle and some field equipment, Tay estuary and Dundee in the background.
Volkswagen Devon caravanette, mobile home and studio; portable canvas hide and collapsible seat; Habicht 30 × 75 close-focus telescope mounted on a tripod; back-up Optolyth 30 × 75 telescope on a car window mount; Leitz trinovid 10 × 40 binoculars; head torch; mosquito net; fish tank and aeration pump.

A head torch helps me to sketch in dark locations and leaves my hands free. This was how I drew the female goosander incubating in her nest in a dark tree hole and also the long-eared bat in a house attic. Standing on a box I was able to get within a foot of the bat and, although it looked at me occasionally in a puzzled sort of way, it was not unduly disturbed.

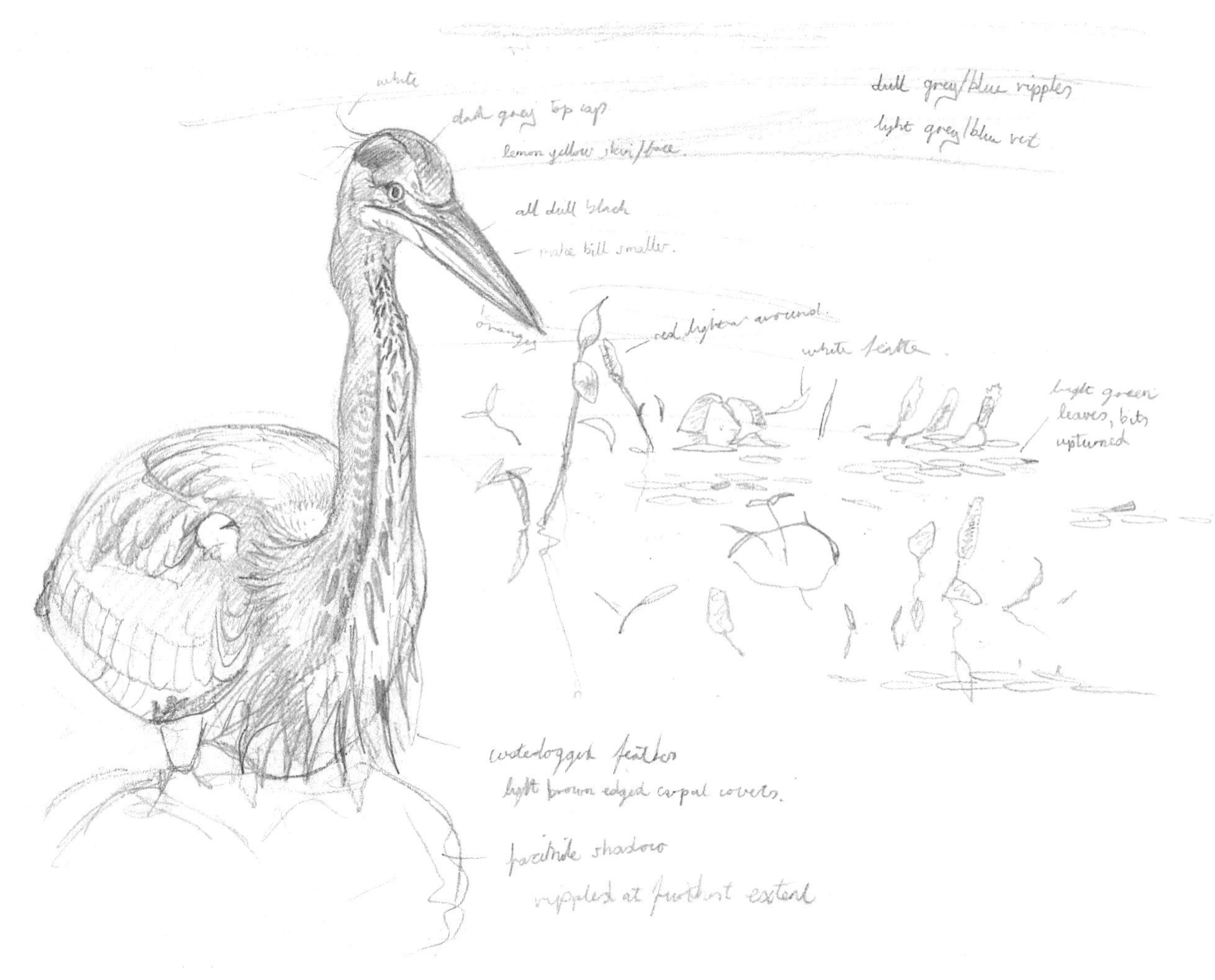

My basic field sketching equipment is itemised on page 18. I keep it to the bare minimum so that I have the lightest load to carry around. The pre-stretched watercolour blocks with various grains of paper are extremely useful in the field. The paper doesn't get wrinkled and I don't have to take separate boards with stretched paper.

The drawing above is what I call a primary sketch, one done quickly on the spot. I put these kinds of drawings in a sketchbook because, through long experience, I can tell that I won't have enough time in the field to work straight onto my watercolour block (this paper is also too expensive to write off if the subject departs prematurely!) In this particular sketch I achieved more than I thought I would with the one pose. Normally I sketch a subject from several angles so as to improve on each continuously as it moves around. I initially sketch the rough outline and form as quickly as I can, adding as much detail as possible while the subject stays near by. Eventually I may get enough details to start a composition. Many of these scrawls don't amount to much more than a few lines but are stored in the sketchbook, and may be a spark for a future painting. Having sketched a particular species I often see just the location for such a composition months or even years later. These sketches, however, quick and simple, have a spontaneity and freshness impossible to emulate in a more detailed work.

Virtually all my initial field sketching is done in pencil, which I find to be the most immediate and sensitive medium. I usually carry a partitioned container with various grades of hard and soft pencils. The more familiar I am with the species the easier it is to record movements – a certain posture or the way a shadow falls on the subject – with only a few lines. I'm drawing what I see, not what I know should be there. It's all about practice. Not everything succeeds in the way I hope, but I'll always learn something new.

Each subject inspires me in different ways. The smew on Loch Faskally was a stroke of luck. I had gone there with the intention of sketching plumage detail. It fed for a while then jumped up onto the fish cage boardwalk to rest, and the juxtaposition with the colourful float was a real and irresistible bonus. Another case of being at the right place at the right time was when I spotted a family party of whooper swans on the ice as I was driving along the road. The abstract patterns on the fresh ice attracted me immediately, as did the deep blue shadows on the birds' plumage with the reflected light from the ice on their underbellies. I painted this directly from the swans as I knew they would remain relatively motionless for a while. White birds present a wonderful canvas for tonal values and shadows uncluttered by feather markings which can sometimes be distracting.

At the same location, several months later, I sketched a pair of coot. The nearest bird had its head tucked away out of sight and each was standing on one leg. The way this blended into one composition fascinated me. It's important not to be prejudiced by preconceived textbook ideas of what the subject should look like. Light transforms colours constantly, and very few people really look deeply at wildlife for any length of time. For instance, a drake tufted duck which may look basically black and white can be transformed into blues, purples, greens and browns at certain angles in strong sunlight. The onset of evening sees yellow and, later, red light influencing plumage, as in the paintings of mallard drakes and the heron at Morton Lochs.

Unlike my two previous books, I have included here very few drawings from dead specimens. One exception is the osprey talons drawn from a bird which had been shot by an aspiring gamekeeper. I could not pass up the opportunity to do a detailed study of its feet which are superbly adapted for catching slippery fish. Another exception is the rough-legged buzzard that was brought to me, which succumbed to starvation overnight. The head was the first thing I painted whilst I could still see the colour of the eye, which darkens, as it dehydrates, soon after death. Later I painted plumage details and its distinctive feathered tarsus, and the result is a lasting reference for future use.

The weather is the final arbiter in what I can achieve day by day. It can also influence me aesthetically as in the roosting cormorants on the old steamer pier near Kenmore on Loch Tay. I sketched the first bird through a slight December snowfall, and became equally interested in depicting the weathered wooden remains of the pier and the rusted bolts. An hour later the atmosphere had changed totally, the sun was shining and a mist was rising from the loch. Another of the cormorants was drying out its wings on a lower perch so I tried to capture that, too, a much more tranquil scene.

I am often asked about my working methods, so I hope this brief description will have given some insight into my approach to my work. I think there is really only one golden rule to follow – enjoy it, regardless of the results!

HEADWATERS

Although the actual source of the Tay originates in the slopes of Ben Lui I have chosen the highest mountain in the Tay basin – Ben Lawers – for most of the mountain flora painting in this book. Ben Lawers reaches a height of 1,214 metres (3,984 ft), and the Lawers range in the Breadalbane region of the Grampians encompasses eight summits over 914 metres (3,000 ft). Much of this area is a national Nature Reserve and is owned by the National Trust for Scotland. A visitor centre open during the summer months houses an exhibition outlining the geological formation and the flora of the area. Ben Lawers is rightly looked upon as a botanist's Mecca because of the rich alpine flora growing on its outcrops of calcareous schist. Some of the species such as alpine gentian (*Gentia nivalis*), alpine forget-me-not (*Myosotis alpestris*), and drooping saxifrage

(*Saxifraga cernau*), grow hardly anywhere else in Britain. And many of these rare plants on Ben Lawers only grow on the vegetated cliff ledges well out of reach of the depredations of that Scottish scourge – the sheep!

Because of the high altitude, the weather affected my work here more than in any other habitat. Many times my sketching trips to Lawers were thwarted by finding the summit shrouded in cloud. What seemed like a beautiful day on the lower slopes would be transformed into a cold windy day on the exposed top. Sitting or standing by a cliff face for hours on end, concentrating on painting a particular plant, makes one realise how susceptible one is to the elements, even on a summer's day. Claire Geddes, in the course of her work mapping the distribution of the alpine gentian, had found a particularly fine specimen, with six flowers, for me to paint. Unfortunately the early August weather decreed otherwise. The alpine gentian usually flowers at the end of July and in the first few days of August and only opens up on sunny days. The weather during that period of 1987 was particularly bad – Claire's tent flysheet was ripped apart one night during a gale – so I had to be content with painting a lesser late-flowering specimen.

The unpredictable mountain weather makes one really appreciate the good days. Probably the finest day of the year was the 22nd of July. The sun scorched down with hardly a breath of wind, and I sat by a ledge near the summit painting the delicate blue of an alpine forget-me-not silhouetted against a dark crevice, thus highlighting the subtle colours. All day at the cliff-face I was kept company by a female ptarmigan. Her well-grown brood of three young were feeding high above me on the vegetated ledges. She talked to her chicks with soft murmuring calls which had a soporofic quality in the still air. Flocks of meadow pipit and wheatear were constantly on the move looking for flies, and a pair of ring ouzel kept their insatiable brood of recently fledged young supplied with food. A kestrel got a good send-off from a cloud of pipits and adult ring ouzels. It was hunting for some of the many voles which travel amidst the rocks and the labyrinth of tunnels. Earlier a party of ravens flew over, their cronking calls reverberating through the hills. They epitomise the free spirit of the mountains.

Working in the hills on such days can be almost a spiritual experience. One feels completely in tune with the natural world. That evening, even though the midges were out in force, I sat near the summit looking over the Breadalbane Hills and beyond, towards a magical vista of hills disappearing miles away into the haze. Lochan a Chait – dwarfed by the amphitheatre of Lawers, An Stuc and Meall G'arbh – was shimmering all over with ripples of rising fish. Loch Tay was far below, and to the south-west the low cloud crept up through the valleys and saddles between the hills like wispy advancing glaciers. The late summer's evening was extraordinarily peaceful. All the noisy groups of 'summit-baggers' had returned to their homes far below, and in their place a group of twenty or so swifts were hawking for flies above the summit a long way from their town nest sites. It had been a day when it felt good to be alive.

After the highly concentrated and emotional activity of painting I need exercise in order to revitalise myself. Fieldwork, especially with birds of prey, fulfills this need, as well as constantly adding to my knowledge. I am part of a

group of enthusiasts in north-east Scotland who monitor the breeding population of birds of prey in the hills, such as golden eagle, peregrine falcon and merlin. This work includes the ringing and wing-tagging of young eagle chicks in order to learn more about them when they leave their natal area. Sightings of the tagged chicks have shown that they leave their parents at the end of October and the beginning of November and travel widely in search of a vacant territory. The peregrine falcon is, happily, still on the increase and we are learning much more about the more elusive merlin population.

I painted all these birds for this book. If a disabled bird is brought to me I can obviously draw it in more intimate terms than I can in the wild. The rough-legged buzzard, for instance, was found just outside the Tay basin and I couldn't pass up the opportunity to paint the details of such a relatively rare migrant from the mountains of Scandinavia. Despite expert veterinary attention from Douglas Brodie the bird died soon after we received it. Weighing barely 500 grams, half its normal weight, the poor buzzard had had very little chance of survival. It was a victim both of exhaustion and of its own inexpertise in hunting – nature's way of population control.

Alpine Forget-me-not, Myosotis alpestris 1/1, growing on a small ledge on a cliff at around 3500 feet, Ben Lawers, 22nd September 1987.

(Alpine) Highland Fleabane,
Erigeron borealis 1/1 Ben Lawers 23rd July 87,
growing on a cliff ledge at 3,500 feet, surrounded by
other plants such as Festuca vivipara, Saxifraga oppositifolia,
Silene acaulis etc

Rock Speedwell, Veronica fruticans 1/1, growing on a rock ledge at around 3,500 feet on Ben Lawers, 20th July 87. Heavy rain has knocked off many of the flower heads.

Drooping Saxifrage
Saxifraga cernua 1/1
Alpine Saxifrage
Saxifraga nivalis 1
Ben Lawers, 24th July '87
both growing on the steep face of a gully by an outcrop of quartzite

Hairy Stonecrop, Sedum villosum 1/1

Sibaldia procumbens

Ben Lawers, August 87, growing at roughly 3200 feet on the wet scree at the base of a cliff

Alpine Gentian ; Gentiana nivalis 1/1 ,
this one growing at 3200 feet amongst a small area
of scree , 6 + 10th August 1987, several tiny
gentians beside the main plant

Purple Saxifrage Saxifraga oppositifolia, growing at 600m's on a mossy bank by a burn in Gleann Taitneach, 18th April 87.

This is one of the earliest hill plants to flower

Holly Fern, Polystichum lonchitis,
growing in a crevice amongst moss
covered rocks, Ben Lawers, 3400 ft
Sept 87

a thick cushion of the lovely lichen – *Thamnolia vermicularis*, near the summit of Ben Lawers, Aug 87

Most of the Ptarmigan were in single sex groups but I saw a couple of pairs together, a fair bit of song-flight display going on.

Pair of Ptarmigan resting amongst lichen encrusted rocks which afford them some camouflage because of the lack of snow, sketched on Meall Odhar, Glen Shee. 6th November 1987. Soon they will loose the grey feathers and become almost totally white for the winter months.

1st year ♂ Merlin, 13th Nov 86   *Falco columbaris*

from an injured bird I am looking after, initially found near Perth.

1st year ♂ Merlin, Dec 86
relaxed and 'fluffed up'

more 'regal'

ruffling its feathers after heavy rain

portraits of my ♂ Golden Eagle, a disabled bird which I've kept for 5 years now, Oct 87.

1st year ♀ Peregrine,
portrait of a wing damaged bird from Glen Garry
which I'm looking after, 30th Dec 87.

Rough-legged Buzzard, Buteo lagopus, details from a first year ♂ found dying from starvation near Pitscottie, Fife on 25th November 87. Despite treatment it died weighing only 520 grammes, it should weigh nearly double this. I couldn't pass up the chance to paint such a relatively rare bird.

26th Nov 87

1st year ♂ Rough-legged Buzzard, 27.11.87
upper tail
measurements - wing 421 mm
bill 21.5 mm
tail 230 mm
weight 520 gms

17
Mountain Hare, Meall Odhar, Glen Shee 6th Nov 87,
sitting basking in the sunshine in a peaty hollow by a
burn, crowberry and heather surrounding it. Almost fully
moulted into its winter coat. Sketched via a telescope some
30 m's distant.

Mountain Hare stretching out in the sunshine,
Meall Odhar, Glen Shee, 6th Nov 87

Lochs

The Tay basin is studded with many lochs of varying size (for my purposes the term 'Lochs' encompasses standing water from flooded fields to the large deep lochs). The largest include Lochs Tay, Earn, an Daimph, Lyon, Rannoch, Errochty and Tummel. They are situated in the upper valleys, and are classed as oligoltrophic lochs, deep, mainly mineral-poor and with relatively poor wildlife. Loch Tay, probably one of the best known of these, is some $14\frac{1}{2}$ miles long with a mean breadth of $\frac{3}{4}$ mile and 155 metres at its deepest. Many of the smaller lochs – largely eutrophic, shallow and mineral-rich – abound with wildlife and, from my sketching point of view, are much easier to get to grips with. One such is Lindores Loch in Fife, a kettle-hole loch formed initially by glacial débris surrounding melting ice. At most times of the year here, there is a wealth of waterfowl, with easy viewpoints from the road. The loch is especially good for diving birds, and I was able to draw an immaculate male smew, a displaying goldeneye and a great crested grebe family via a telescope from my vehicle.

Morton Lochs, a National Nature Reserve near Tayport, are artificially created lochs. Originally excavated as fish-rearing ponds they have now been landscaped into an attractive wetland habitat. The shallow lochs usually dry out somewhat in the late summer providing exposed mud which attracts waders on passage, such as greenshank, spotted redshank, green sandpiper and ruff, to name but a few. There is a public hide and one on stilts overlooking the loch available to permit holders. I spent many entertaining hours in the latter sketching herons, mallard, black-necked and little grebes fishing together, and a spotted redshank.

These smaller water bodies are far superior in wildlife to the huge lochs; indeed a flooded area in a field can be a great attraction to many species. One such treasure trove is a series of pools by Kinkell Bridge near Auchterader. Their position right beside the road enabled me to sketch comfortably from my vehicle without disturbing the wildlife. An unexpected bonus here was a vagrant male green-winged teal. This is the Nearctic equivalent of our own teal and is only rarely recorded in Britain (almost all the records that have been made are of males because the females are impossible to tell apart). The American male differs from the European mainly by having a white crescent between the breast and flanks and by not having a white lateral stripe above the black one along the wing edge. I had superb views of this individual as it dabbled along the edge of the pool at ranges down to ten metres. Other species of duck present here from time to time include mallard, wigeon, gadwall, shoveller and teal.

Waders can also be found in good numbers from the late summer through the autumn. These included such birds as the lapwing, snipe, curlew, oystercatcher, common sandpiper, redshank and greenshank. On two consecutive days I painted a snipe and a lapwing. The snipe was mostly hidden in the lush grass, resting between bouts of probing deep in the ground for worms. A very obliging juvenile lapwing fed close to the van allowing me to get enough detail in a diffuse yellow light. A juvenile heron also spent quite a lot of time at the pool, often wading right up to its belly amongst the amphibious bistort.

On another loch a heron provided me with a chance to paint a bird's interaction with man. It was looking down forlornly on a milling mass of fish in a cage netted over to prevent any access by predators. The owner of this fish farm had a refreshingly tolerant attitude to the herons. They caused him very little harm apart from spearing a few fish through the side netting. A black bin behind the heron provided a regulated supply of fish pellets for the fish below.

Trout fishing is one of the reasons for the decline of the black-throated diver, a species very susceptible to disturbance during the breeding season. It is now down to only a few breeding pairs in the Tay area. As their nests are on the banks of lochs or islands they are sometimes disturbed by people who unwittingly keep them off their eggs for long periods. Fluctuating water levels do not help much either; because the diver is so ungainly on land its nest has to be very close to the water's edge. I sketched my divers only when they had chicks, and from a distance so as not to disturb them too much. The diver's beautiful sleek shape and striking plumage, and its black velvety throat shining deep purple in strong sunlight, attract me very much.

Another favourite of mine is the whooper swan. To hear its magical trumpeting calls from a misty loch is one of the real joys of nature. I painted a portrait from an injured swan which I found sitting amongst sheep in a grass field. It is normally a gregarious bird, so when I walked out into the field I fully expected it to fly off. It tried to do so but all it could do was stagger around flapping its wings in vain. After subduing the swan I took it to a vet, Douglas Brodie, to ascertain what was wrong with it. X-rays and blood samples showed no visible signs of damage. Actually it caused Douglas's assistants more harm because it kept beating its wings and snaking its head and long neck around to grab their ears and hair in its serrated bill! They were relieved to see the back of this ungrateful patient. The swan gradually recovered the use of its legs during a period of convalescence amongst my own captive wildfowl. Then one day it took off into the blue yonder, free again at last. I can only surmise that it had had a temporary spinal injury resulting from a collision with wires.

Some of my work has a rather tenuous connection with water, but what is artist's licence, after all? For instance, I simply had to include the blackcock because I was passing a new lek of young colonising males and couldn't resist painting them as they sat dejectedly in the pouring rain. Comfortably ensconced in my camper with the rain pounding off the roof I could fully sympathise with

them. Their numbers are increasing in Tayside, benefiting in the short term from the expansion of the newly planted Sitka forests.

Despite its majesty in the air and its spectacular plunge when diving for fish, the osprey at rest is a dishevelled, loose-plumaged bird. Ospreys spend much of the day perched near the nest containing their large chicks. While they are inactive I have plenty of opportunities to draw them at a distance. The detail of their feet, drawn from a dead bird, shows how well adapted they are to catching fish. Spiky undersoles and long needle-sharp talons help to grip the slippery prey. In addition the osprey's outer toe is reversible, enabling it to grip the fish with two toes on either side.

As well as drawing ospreys I do as much as I can to help conserve them. Since 1982 Steve Cooper and I have been monitoring the increasing osprey population within the Tay basin. One nesting pair can usually be easily viewed from the hide at the Loch of the Lowes near Dunkeld run by the Scottish Wildlife Trust. This helps to alleviate unnecessary disturbance at more confidential sites which can be prone to excessive visiting. During 1987 one such pair deserted their nest because they were so disturbed by the presence of onlookers, who included one well-known Scottish climber, who should have known better; he sat attempting to photograph fishing adults near the nest adjacent to a small lochan.

Egg collecting is still a problem, a sad relic of the Victorian era. Two of our nine osprey clutches were stolen during 1987. This was a serious setback to their recolonisation of Scotland as the ospreys seldom relay; thus that particular year's production for the three aforementioned pairs was lost. To combat situations like this we sometimes have to take drastic action. For example, we had to fell one easily accessible nest. This osprey eyrie had been robbed the previous year so we built another eyrie for them some distance away in a well 'doctored' tree. To discourage unwanted visitors we cut most of the main branches off flush with the trunk, wound barbed wire round the top and applied liberal quantitites of grease to prevent anyone shinning up the tree. Conservation in action!

Building man-made nests in new areas or where birds have been seen helps to establish new pairs. One such nest was built by Mick Marquiss and myself to replace a new eyrie, from an inexperienced pair, which had blown out in a gale, smashing the eggs. We sawed the top off a huge Douglas fir to get down to a good radial base of branches. Then we fixed in an old wire tattie basket to form the nest cup and surrounded this with a web of thick fence wire. This flexible construction is much better in high winds than a nailed framework. To this we tied in a framework of large branches which we interweaved with smaller twigs to form a large nest nearly 100 cms in diameter. Just before the adults were due back from their wintering grounds in Africa we lined the new nests with dried vegetation, then spread grass and moss round the nest cup to form a comfortable base to the eyrie. As a last touch we sprayed the outer edge of the nest with white paint to make it look well-used and more noticeable from the air. To our delight the ospreys occupied this nest straight away and raised a brood of two, a great reward for all our effort.

A man-made nest containing two 26-day-old osprey chicks, an infertile egg and the remains of a perch.

Most of the chicks are colour-ringed as part of a scheme run by Roy Dennis, the Highland officer for the Royal Society for the Protection of Birds, to monitor the future colonisation spread and to follow individual life histories. These plastic rings have a different colour for each year and the individual markings etched on each one can be identified with a telescope from a fair distance. The age of first breeding is usually three years and three eggs is the usual clutch from established pairs. To discourage egg thieves we disfigure the lovely egg markings with an indelible pen. In 1985 I ringed and recorded the first brood of four chicks to fledge in Scotland (one clutch of four had previously been recorded in the Highlands but had been stolen by an egg collector).

The ospreys fish on the rivers as well as on the many fish-stocked lochs in the area. A preliminary analysis of fish remains collected by us and identified by Mick Marquiss shows that the ospreys' diet consists of 42% trout, mostly rainbow with a maximum length of 41 cms; 29% perch, maximum length 33 cms; 18% roach, maximum length 31 cms; and 11% pike, maximum length 40 cms. The chicks are very docile in the nest in comparison with many raptor chicks and usually lie prostrate in the nest cup. When we are ringing chicks the adults generally keep at a distance, calling in alarm, though there are exceptions and some of them can and do strike at an intruder. One adult osprey made a few stoops at me coming within a metre before sweeping away. The hair on the back of my neck stood on end!

But that sort of hazard is a small price to pay for the prospect of helping birds such as ospreys. All those long hours of fieldwork are repaying only in a small way the huge debt I owe to the wildlife I draw and paint.

Above, a 20-day-old osprey chick. The colour ring is covered by sellotape to hold the ring together whilst the fixing glue dries.

Below, the first recorded brood of four chicks to fledge in Scotland.

a pair of toads in their 'mating embrace' from a pond by my house,

Dron, 19th April 87

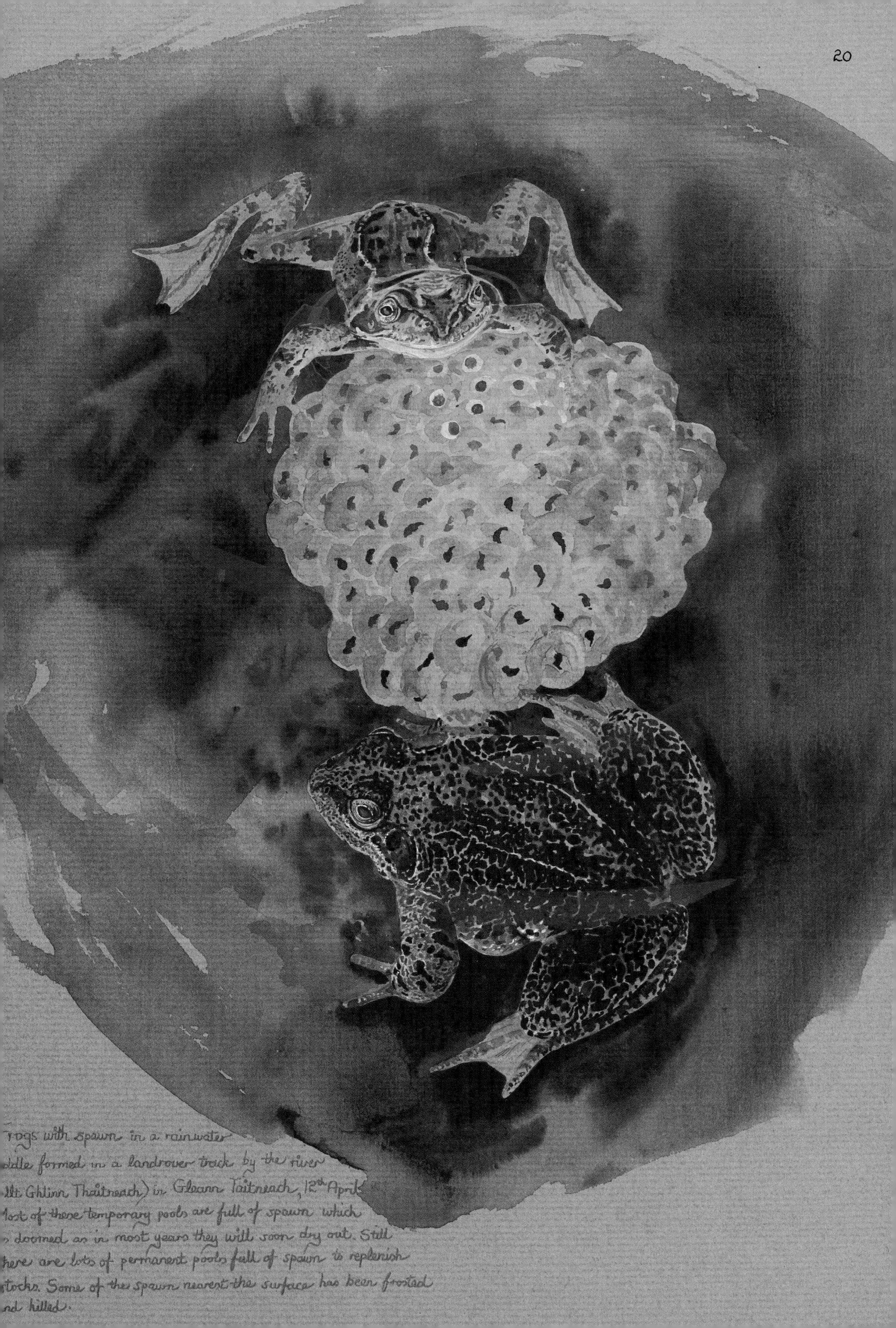

rogs with spawn in a rainwater
ddle formed in a landrover track by the river
llt Ghlinn Thaitneach) in Gleann Taitneach, 12th April
Most of these temporary pools are full of spawn which
doomed as in most years they will soon dry out. Still
here are lots of permanent pools full of spawn to replenish
stocks. Some of the spawn nearest the surface has been frosted
nd killed.

young Roebuck, feeding by the shore of Loch Moraig, Blair Atholl 12th April 87

Blackcock on their lekking ground in the pouring rain near Aberfeldy, 18.00hrs, 28th May 87. The aggressive lekking display is much less intensive than a few weeks ago, these are younger ♂♂ in a newly colonised area.

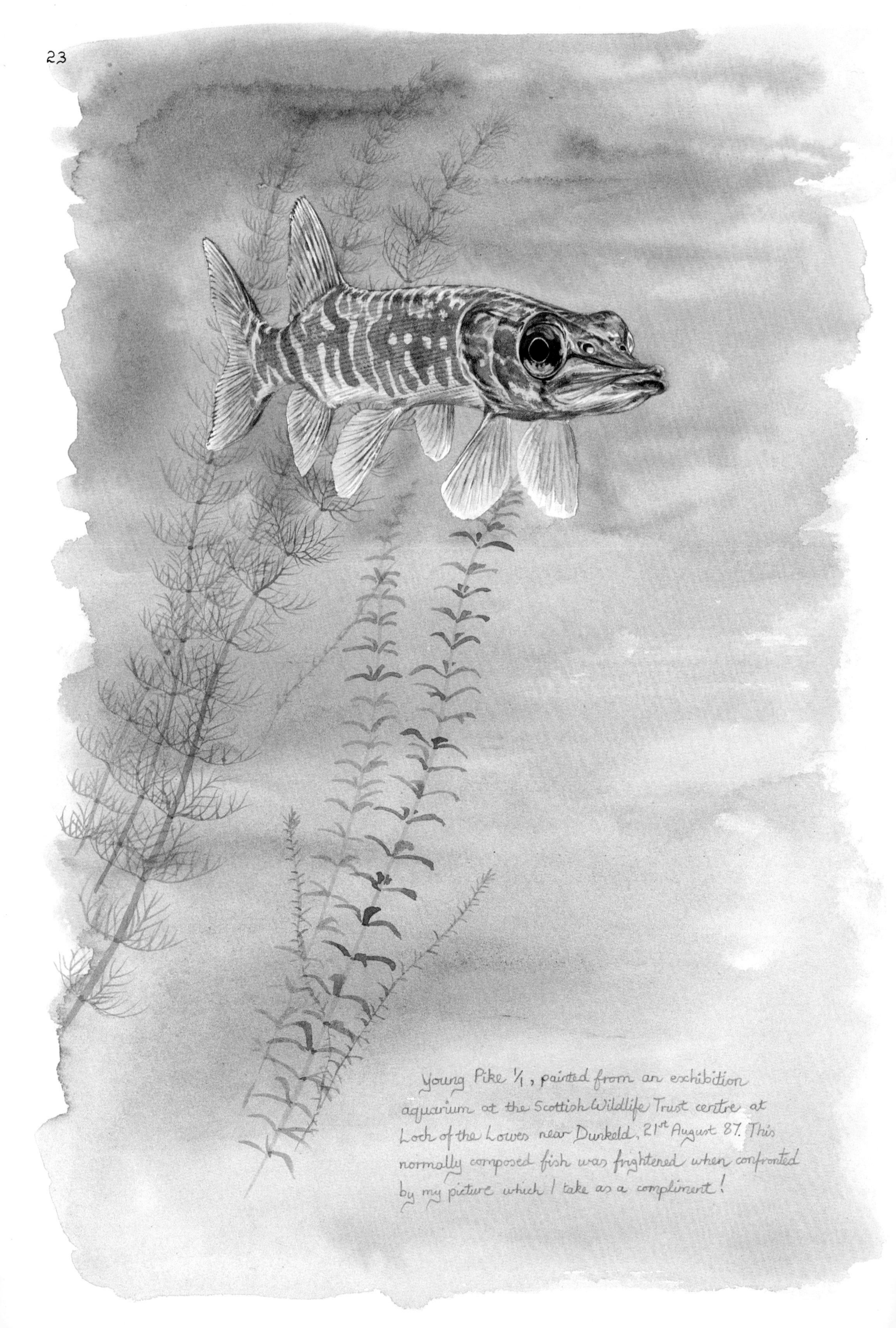
Young Pike 1/1, painted from an exhibition aquarium at the Scottish Wildlife Trust centre at Loch of the Lowes near Dunkeld, 21st August 87. This normally composed fish was frightened when confronted by my picture which I take as a compliment!

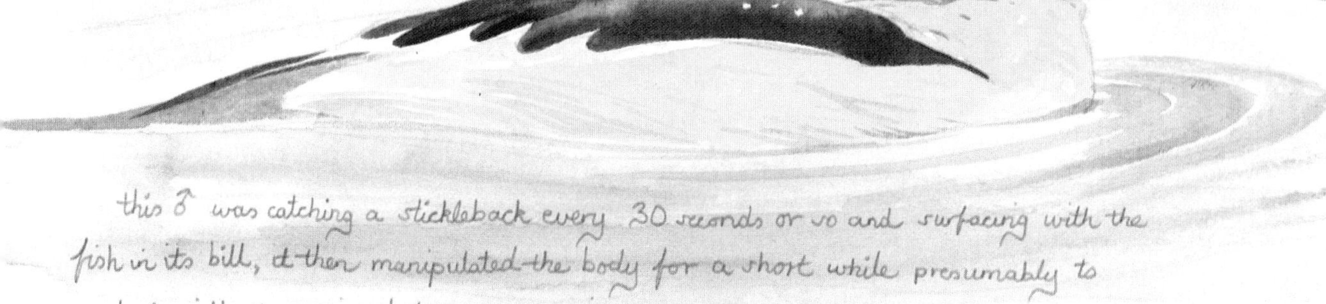

this ♂ was catching a stickleback every 30 seconds or so and surfacing with the fish in its bill, it then manipulated the body for a short while presumably to deal with the spines before swallowing the catch

Goosander sketches, Drumore Loch, 10th April 87

♂ Osprey, near Dunkeld
21st July 1988

creamy shading
dead
Norway Spruce
♂ Osprey, nr Dunkeld, 18.7.86

This male osprey came in with a rainbow trout and landed on top of a dead tree adjacent to the eyrie. He proceeded to eat the fish while seeming oblivious to the two large chicks in the nest below who looked up longingly and hungrily. Then he fluffed himself up and sat for over an hour while I drew him. In the meantime the chicks sank dejectedly back into the nest cup. The female was sitting about 50 metres away in another tree.

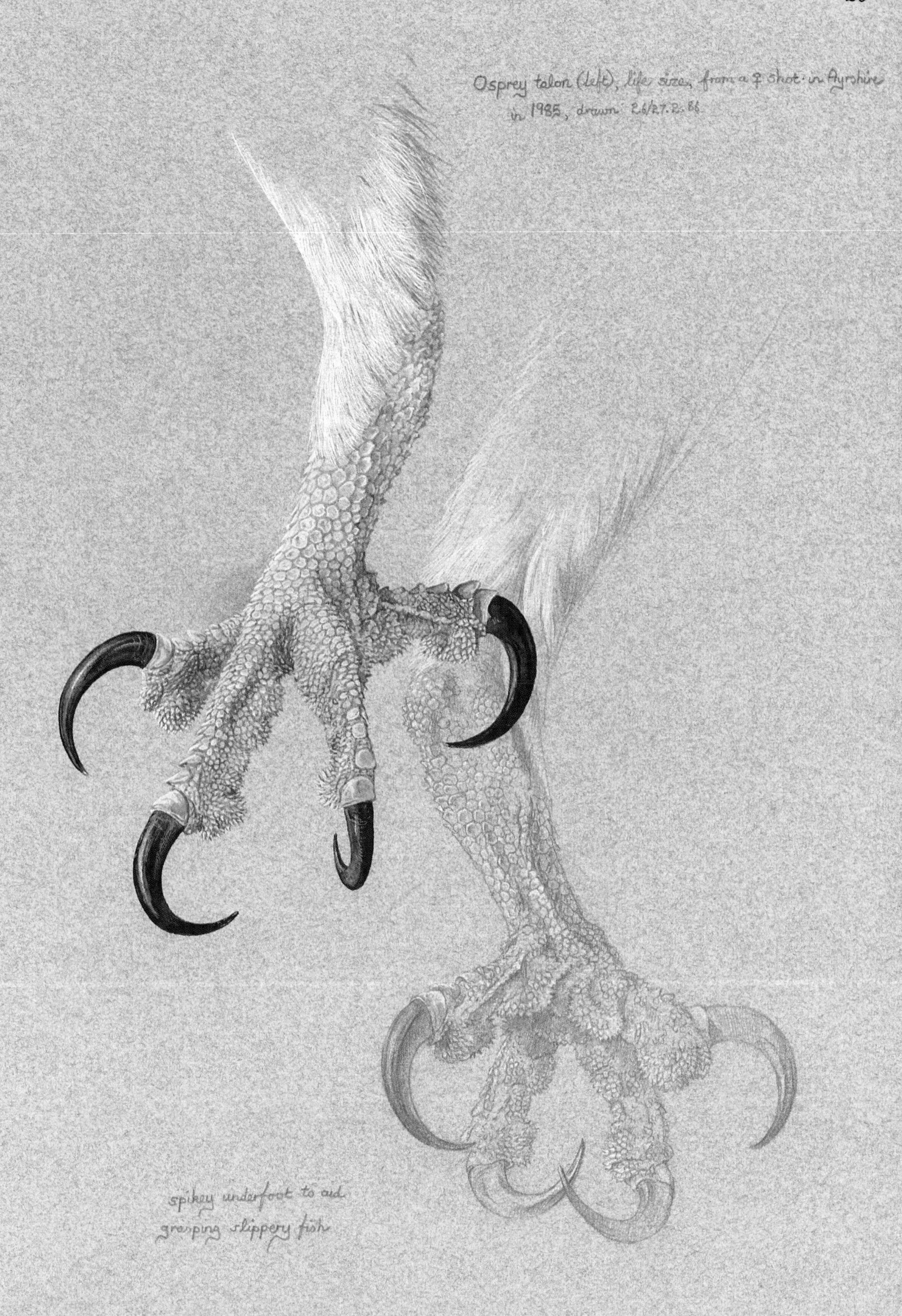
Osprey talon (left), life size, from a ♀ shot in Ayrshire
in 1985, drawn 26/27.2.86
spikey underfoot to aid
grasping slippery fish

Nest liberally lined with moss, grass, heather, lumps of peat and large pieces of bark and rotting wood

Osprey chick (c 40 days old), 11th July 87
These must be the least aggressive of all raptor chicks, this chick hardly moved at all whilst I sketched it in pencil for 40 minutes after ringing it. They usually lie prostrate like this when disturbed. I painted this later today with the help of colour notes and a few bits of the nest lining.

♀ Osprey at her empty eyrie, bird
sketched 6th August 87, nest on a dead Norway Spruce

juvenile

preening

Black-throated Divers, Perthshire 14th Aug 86

the black throat has a purple gloss in the sunlight

A pair of Black-throated Divers with their week old chick in the rain, sketched from a distance via a telescope, 10th June 87

34

sketches of a ♂ Smew
Lindores Loch, Fife
16th April 87

1st year ♂ Smew,
Loch Faskally, Pitlochry
23rd March 87 (present since at least 11th January)
op- reflections of spruce, birch & bracken.
resting on the boardwalk
surrounding a fish cage

'On Ice', Whooper Swans resting on Drumore Loch, Glen Shee, pair plus their youngster. 21st March 87

Whooper Swan, portrait of an injured bird on my garden's pond, 13th Dec 87,
this bird hasn't regained full use of its legs as yet after hitting wires which resulted in a spinal injury.

Common Gull on its nest on an old fallen Norway Spruce, Loch Moraig, Blair Atholl
(some nests were 50 feet up in Scots Pines)
19th May 87

♀ Mute Swan on her nest,
Stare Dam near Dunkeld,
29th May 87

Black-headed Gulls and a ♂ Tufted Duck,
Red Myre, Fife 16th June 87

bathing

mare's tail
Hippuris vulgaris

Pair of Coot, one seemingly headless with its head tucked under a wing, resting on a rock, Drumore Loch 3rd August 87.

Great-crested Grebes, adults and chicks, Lindores Loch.
17th August 87

Adult Little Grebe and juvenile Black-necked Grebe
between dives, Morton Lochs 18th Aug 87
44

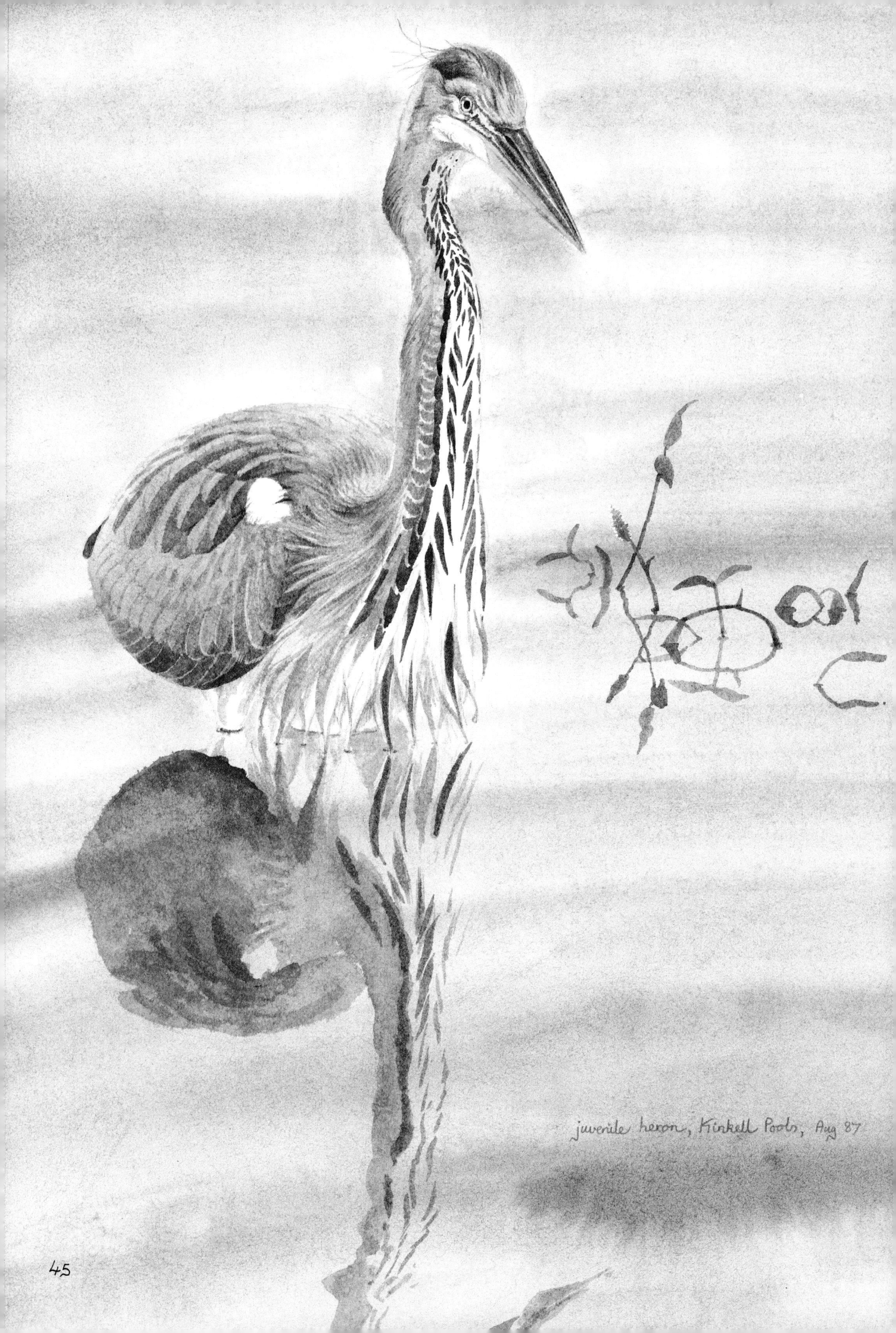

juvenile heron, Kinkell Pools, Aug 87

2nd yr Heron sitting on fish cages, Drumore Loch, 16th May 87,
the nets over the cages preventing any predator access. The
contains fish food pellets

Juvenile Heron hunting at sunset, Morton Lochs, Fife
16th Sept 87

Drake Mallards moulting out of their 'eclipse' plumage,
Morton Lochs 19.00hrs, 16th Sept 87

Drake Mallards resting on ice, Kirkell Pools Auchterarder
3rd Dec

♂ American Green-winged Teal, Anas crecca carolinensis,
sketched at ranges down to 30 feet in a flooded field corner by Kinkell
Bridge, nr Auchterarder, 6th April 87 - dull, overcast & drizzle. This ♂
was slightly larger than the nominate race with finer vermiculations
on the grey body feathers, dabbling in the shallow edges.

displaying Goldeneye, Lindores Loch, Fife
24th February 87
thin ice in foreground, lots of gull feathers around.

adult ♀ Goldeneye, 31st Dec 87
from one of my captive ducks.

♀ Goldeneye on my pond with ice
February 86
54

Snipe at rest amongst the grass in a flooded field,
Kinkell, By Auchterarder, 4th Oct 87

1st year Lapwing feeding in the flooded field at [illegible]
5th October [illegible], yellow overcast light.

juvenile Spotted Redshank, at dusk, Morton Lochs, Fife
25th August 87

Pink-footed Goose in the snow
on my pond, February 87

1st year Cormorant on the remains of the old 'steamer' pier
near Kenmore, Loch Tay, 8th Dec 87 (11 Cormorants present)

An hour later the snow had cleared with mist
rising, old pier nr Kenmore, Loch Tay, 8th Dec 87

Dipper and an unusually marked
♂ Goldeneye (Smew hybrid?) on Loch Tay
Dec 87.

floating on the surface between dives

Dipper, shore of Loch Tay, 8th Dec 87
(they sing beautifully at this time of year)

Dipper nest on an alder branch overhanging
the River Braan, near Trochrie, Dunkeld, Oct 87.
An unusual nest site but safe from predatory mink.

MAIN RIVERS

The River Tay is the confluence of many rivers and their tributaries draining the largest catchment area in Britain. These rivers pass through some of the most varied and beautiful scenery in Scotland and are famous for their salmon fishing. The fishing is split into many beats by the riparian owners and brings considerable tourist income to the area. Virtually every good pool and landmark on the river has a special name, with its own folklore; some, to conjure with, include the Tail of Clachantaggert, Cat's Condie, Galligan's Croy and Ferny Haugh.

I did most of my fish work on the River Earn which enters the Tay some seven miles below Perth. It was a more manageable size and had particular advantages. Allan Allison and his friends regularly fished a beat here at Kinkell Bridge, the ghillie Tom Adamson was a great help, and through them I could easily find out about river conditions and the likelihood of catching fish. There was also a good fishing hut where I could paint in comfort on cold spring and autumn days. Listening to the fishermen's banter during meal breaks was fascinating. I had many a silent chuckle as I painted away in the background hearing tales of epic struggles with salmon and the inevitable 'one that got away'. As the whisky took effect both the noise level and the size of fish caught increased in proportion to each other! Emotions ranged from sheer elation at a good catch, with many fish running, to deep despondency after flogging the river all day with no sign of fish at all.

Whilst waiting to catch fish I busied myself with various ploys. But I failed to draw that flashing blue jewel of the river, the kingfisher. Although I often saw them flying up and down the river I never got to grips with a stationary one with telescope and sketchbook at hand. Just below the beat was a sand martin colony with some 76 occupied burrows in the sandy riverbank. Unfortunately most of the second broods were drowned when the river rose substantially after heavy rain. Despite such setbacks the numbers of sand martins in the area are increasing after a disastrous drop in the early 1980s, which occurred because drought increased the area of the Sahel Desert on the sand martins' migration route through Africa.

People seldom appreciate the beautiful colours of fish. I vividly remember one fisherman's wife peering over my shoulder at one of my paintings and remarking, 'they aren't that colour'. Politely I asked her to look closely at the

Above, a heron clutch. Below, tiny heron chicks.

fish and convinced her that salmon are not just silver-coloured. They vary considerably in colour according to the season. Fresh-run fish in from the sea with lice attached do indeed appear like a bar of silver; the sea lice indicate that the salmon has spent no more than 36 to 48 hours in fresh water because after this period the lice fall off. But when the fish have been in the river for a while they colour up and become more reddish as spawning time approaches. It is lovely to watch these great fish leaping out of the water, an iridescent bar catching the light before plunging back into the river. On the Almond there is a beautiful waterfall known as Buchanty Spout. With a good water flow this is a spectacular place from which to watch the salmon trying to ascend the falls. It attracts many visitors and one can stand on the rocks barely a few metres from the leaping fish. My oil painting of leaping salmon was based on this waterfall.

Herons are a favourite bird of mine and I never tire of painting them. In fact I consciously had to forbid myself from sketching any more for the book. They are full of character, and seem to pop up everywhere in eminently suitable locations. They have a beautiful breeding plumage, a bright orange-red bill, a subtle lilac neck colour and a spray of long white plumes at the base of the neck. I had great fun watching and sketching them from the heronry hide.

Building this hide involved a lot of effort, mostly in constructing a temporary platform some eleven metres up in the trees. Steve Cooper and I used three five-metre poles to form a triangular base spanning a ride just uphill from a couple of heron nests in a Norway spruce plantation. Using one heavily branched tree on the ride edge as a ladder we hauled up the poles and lashed them together across the ride to our three chosen trunks. This was done in stages so as not to disturb the birds too much in the latter days of incubation. Once the base was fixed to the main tree trunks I nailed the cross poles at their intersections and fixed a wooden potato pallet on top as the floor. This had been pre-drilled to take the hide poles. A few days later I placed the canvas hide on the platform only half upright. The next day, once the birds were used to this, I erected the hide fully and guyed the poles to nearby branches to keep it taut. By this time the nearest clutch had started to hatch.

When I entered the hide some of the nearest adults would fly off, but they usually returned within a few minutes encouraged by the incubating birds, out of sight of the hide, which were not disturbed. The hide was really too close to use the telescope, except for portraits, as my field of view was too restricted. Instead I mostly used my binoculars taped onto the tripod pan-head. The only problem was a pair of jackdaws which insisted on building a nest inside the canvas hide. At first I thought a friend had been playing a trick on me by putting an old heron's nest inside the hide. When I went into the hide early each day I had to eject a mass of birch twigs and sheep's wool which the industrious jackdaws had manoeuvred under the hide door flap. Eventually they managed to build a big enough nest, between my visits, to lay two eggs! These had to be ignominiously thrown out as well and, thankfully, at long last, the jackdaws gave up their struggle and moved elsewhere.

The goosander and red-breasted merganser, collectively known as 'sawbills' because of the serrated edges to their bills, are fascinating birds. They are still much persecuted (though a licence is required for shooting them) throughout the river system for the heinous crime of taking salmon parr. So far there is little scientific data either to prove or disprove if this does any lasting damage to fish stocks: considering that a salmon lays between 2,000 and 15,000 eggs, only a tiny percentage need survive all the dangers to perpetuate the species. Still, despite this shooting the birds appear to be holding their own. The drake goosander is at his best in December and January. His body plumage is a deep orange-pink which gradually fades to white by the early summer. One of the easiest places to see them is from the riverside pavements in Perth itself.

I sketched a nice female on her nest by the River Almond. Situated at a convenient height in the hollow trunk of an old gnarled alder, she hissed at me like an angry snake. Surely this aggressive display would put most would-be predators off? Indeed, as I was sketching her, a jet-black mink came by, hunting along the opposite river bank. Not renowned for its eyesight, and being downwind, the mink didn't see me and continued on its way.

The plumage of the male red-breasted merganser is altogether more complicated than that of his clean-cut cousin. A smaller bird, he has a spray of spiky feathers on his head and neck, a white neck ring, spotted buff breast and grey vermiculated flanks. On a shingle island just below Dunkeld bridge, I managed to sketch both species together. The females look fairly similar but can usually be told apart by the shape of their head and crest. The female merganser has a more diffuse transition between its brown head and greyish neck compared with the goosander's sharply delineated head and white throat patch. Normally more common on the lower reaches, the merganser appears to be colonising the upper river in recent years. Unlike the goosander it usually nests on the ground amongst vegetation, often on small islands. It also lays some five weeks later than the goosander.

Between Perth and Pitlochry there are many shingle banks bordering the river. These and the adjacent grassland are fertile areas for plant and bird life. I concentrated on an area near Dalguise which was easy to get to and had a good variety of breeding birds. Here I spent many hours in my hide drawing various species on the nest. The aptly named common gull was the most abundant species, nesting amongst the grass, stark white against the green. Focusing on their heads I explored their form with soft blue shadows. Without moving the hide I was also able to sketch an oystercatcher incubating its clutch nearby in the shingle. Similarly, during my work on a nesting common tern, I watched an excited pair of common sandpipers displaying close to the hide. They were prospecting for nest sites and showed particular interest in a bramble bush some three metres from the hide. To my delight the sandpiper eventually laid in this site. I was able to view the incubating bird through a gap in the vegetation. The strong shadow and its ruffled plumage were aesthetically delightful, and the prevailing air of calm was unusual for such a frenetic species.

Closer to the river a ringed plover stood sentinel to its mate secreted amongst the shingle. Even though I knew where the clutch lay I found it nearly impossible to spot the incubating bird. Their bold facial markings blended in remarkably well with their surroundings. Later I was fortunate to find their four young chicks crouching motionless amongst the stones, an instinctive survival strategy at the first sign of danger. Two of the chicks fledged successfully, I'm glad to say.

Small White Orchid 1/1
Leucorchis albida, 17th June 87
Scottish Wildlife Trust reserve at
Keltneyburn, near Aberfeldy

Greater Butterfly Orchid, Platanthera chlorantha 1/1
Keltneyburn, Aberfeldy 8th July 1987

Fragrant Orchid, Gymnadenia conopsea 1/1,
a bit past its best, aptly named with it's wonderful scent,
Keltneyburn, Aberfeldy 8/9th July 87

Common Spotted Orchid, Dactylorhiza fuchsii 1/1
with Ragged Robin and Buttercups.
Keltneyburn, Aberfeldy 12th July 87

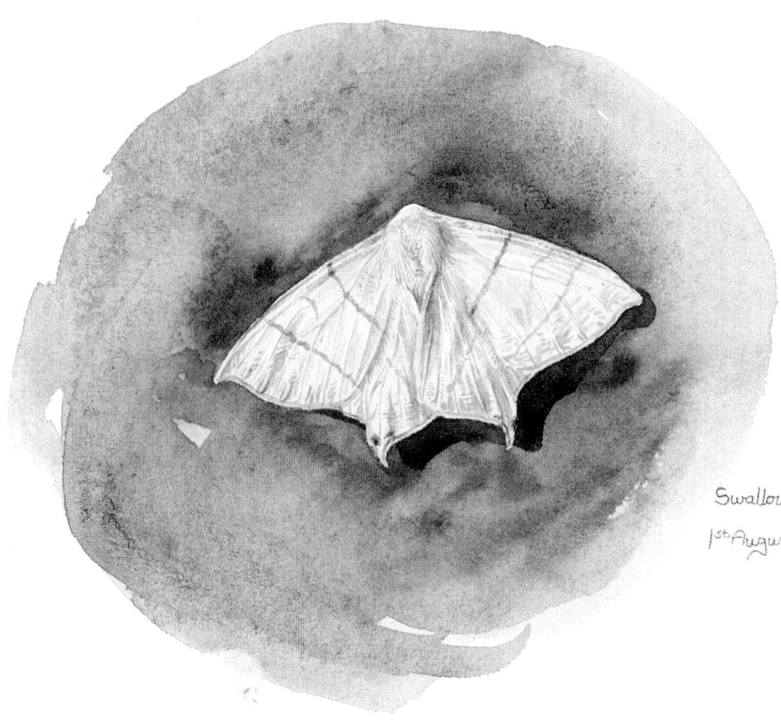

Swallow-tailed Moth. Ourapteryx sambucaria 1/1
1st August, attracted to the spotlight on my duckpond.

Garden Tiger, Arctia caja 1/1
2nd August. 87

Burnished Brass, Diachrysia chrysitis 1/1
3rd Aug

Antler Moth. Cerapteryx graminis 1/1
19th August

Long-eared Bat, roosting in the rafters of a Perth house (Heasman's), sketched with the aid of a head torch, 2nd Sept. 87, looking directly up at the bat.

69

dappled light on fur & stems

pinkish bare skin

white whiskers

new rasp shoots

Hare, ½ asleep in shade of raspberries, 15.00hrs
Millhill 7/5/85

white underbelly, diffuse orange border

Some sketches linger in the mind as a possible painting for the future. One such sketch was this brown hare lying soaking up the sun in a raspberry field. It may look scruffy but this was due to the dappled shadows and its fluffed up fur. His eye was half open, as he kept watch on me warily while I sketched it from the roadside. Tayside has long been famed as a raspberry-growing area.

© Keith Brockie

♀
Roebuck, Dron, 7th February 87
09.00 → 11.30 hrs
this buck is in his 3rd year.
rufous
(also below ear)
lying down in the long grass
by the burn

Sketches from a 9½ lb ♀ Salmon fresh upriver from the sea,
caught by Peter Kelly, Kirkell Bridge, Earn - 15th September 87
below, a sea louse attached to another fish showing how fresh the fish are - these parasites usually only stay on for 36 to 48 hours in fresh water
anal fin

74
River Earn, Kinkell (W.Wilson 10½ lb)
© Keith Brockie Oct 1986

Head detail of a 5½ lb fresh run ♀ Salmon, 1/1,
caught by Tony Oliphant on the Earn at Kirkell Bridge,
17th September 87

'Portrait' of an ugly 42 lb cock Salmon, 2/3 life size, caught on the Scone beat of the Tay by T. Johnston of Coventry, drawn 28.9.86. At this time of year close to spawning the male's head is misshapen and the lower jaw has become 'hooked' - known as the 'kype'. The growth rings on the scales under magnification show that this fish has spent 2 years in fresh water and 3 years in salt water.

77

detail from a 7½ lb cock salmon caught on the Earn by Vernon Burke
14th September 1987

An adult Great Black-backed Gull feeding on salmon which have died after spawning (kelts). Other birds joining in this feast included Herring Gull, Black-headed Gull, Waterhen, Coot, Heron, Mallard, Carrion Crow and Robin!

Sea Trout, roughly life size
Kinkell Bridge, River Earn 4th & 5th March 1987

The upper fish of 1½ lbs (caught by B. Marshall) was a fresh run fish from the sea contrasting with the lower fish (caught by A. Allison) which has been in the river for a while. The smaller fish, called a Finnock, is about 18 months old. They were painted live in a tank and lack the lustre of a fish out of water.

A small Grayling 1/1, from River Earn by Kinkell Bridge.
2nd September 1987

♀ Mallard on her nest hidden amongst dead bracken by the burn just below my house, Dron. 15th May 87. Her beady eye watching me as she flattened herself to become even more inconspicuous.

15th April 1987, my first sketch from the heron hide showing my view via the telescope of an adult incubating it's clutch of eggs. What marvellous intense eyes these birds have.

View from the heron hide
platform, near Aberfeldy
27th May 87

adult Heron at its nest, the head looks
big due to the fluffed up feathers
nr Aberfeldy, 25th April 87

Heron brooding tiny young, painted in the hide via a telescope, near Aberfeldy 16th April 87

adult heron dozing on the edge of the nest whilst guarding its small chicks
N. Aberfeldy, 25th April 87

Adult Heron guarding its small chicks
Nr Aberfeldy, 26th April 87

raised crown f.

Heron chicks panting in the sunlight
nr Aberfeldy 23/6/87

nearly fledged heron chick,
nr Aberfeldy, 23rd May 87

well grown Heron chick c37 days old, the other two
larger chicks were clambering about in the branches above
the nest, nr Aberfeldy, 23rd May 87

Herons preening & sunning themselves
nr Aberfeldy, 25th April 87

♀ Goosander, sketched for 20 minutes initially with the aid of a head torch. She was hissing at me like an angry snake, Sma' Glen, 22nd April 87

Goosander nest site in the hollow trunk of a stunted
old alder tree by the River Almond above the Sma' Glen,
24th April 87. The entrance to the hole was one metre above
ground level. she had ten eggs on the 20th when I visited the
site to draw her but she was off feeding.

pair of Goosander and a ♂ Red-breasted Merganser resting on an island just below Dunkeld Bridge on the Tay, 20th April 1987

♂ Goosander; Tay in Perth just below Queen's Bridge,
fishing in the strong spate currents, 5th January 87

incubating Common Tern,
on shingle bank by the Tay at Dalguise, 28th May 87
'flat overcast light'
½ asleep
This pair had 3 eggs a week ago but are
now down to a single egg due to some predator

Common Gull portraits from a nesting colony on the banks of the Tay by Dalguise, 7th May 87.
Lovely shadows in the strong sunlight.
asleep brooding
½ asleep on guard duty!

♀ Lapwing on her nest, Tomchulan, Glen Brerachan near Pitlochry
17th April 87

sketched from my camper at the roadside via a telescope.

Incubating Common Sandpiper, sketched from a hide some 8 m's away, ½ asleep and fluffed up with the sun casting shadows over the bird. The nest was under a bramble bush on the shingle banks of the Tay at Dalguise, Nr Dunkeld, 20th June 87.

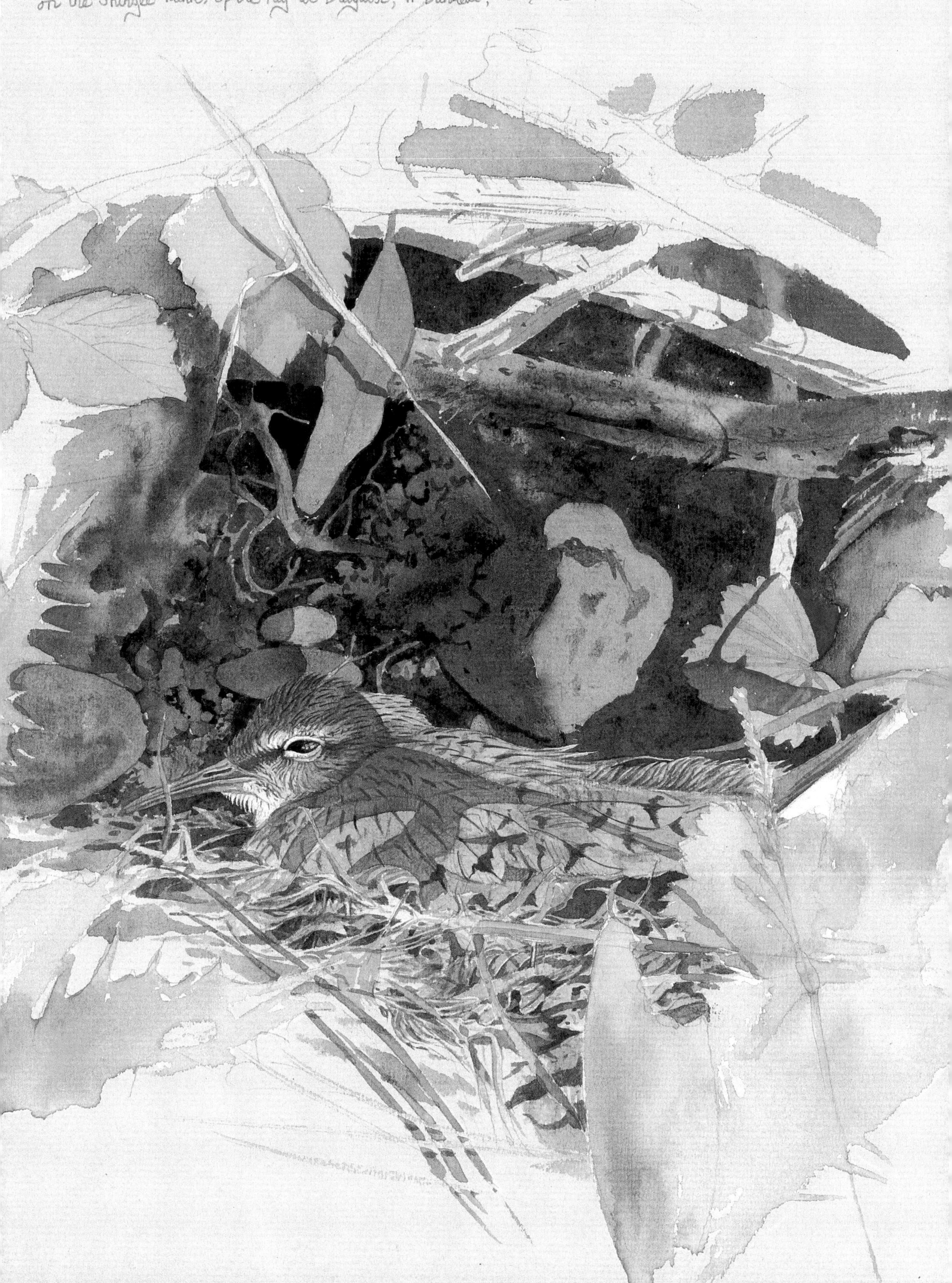

droplets of water on the bird and stones after a slight shower
♀ Ringed Plover on her clutch of 4 eggs (3 yesterday) amongst the shingle on the banks of the Tay at Dalguise. Sketched from a temporary hide some 12m's away.
1st June 87

3 of the four Ringed Plover chicks hiding amongst the shingle on the banks of the Tay at Dalguise, 1st July 87. These are 5 - 6 days old, stones drawn later.

one of the chicks on the 16th July, head detail from a bird in the hand

Oystercatcher incubating it's clutch amongst the shingle
on the banks of the Tay by Dalguise, 7th May 87
Sketched from a temporary hide via a telescope.

Oystercatcher brooding chicks during a shower on the shingle by the Shee Water, by Spittal of Glenshee, 14th June 87

Sand Martin colony (part of) in a bank near the rive
Tay near Aberfeldy, 7th July 87

Young Sand Martins at the entrance to their nest tunnel on the steep sandy bank of the River Earn near Kinkell Bridge. Most of the first broods are already on the wing, these ones are eying the outside world and waiting to be fed, 29th June 87.

Sadly in August a large rainfall flooded the river and all the second broods were drowned, some 76 burrows in all.

♀ Grey Wagtail flycatching over the stream in my garden, in tail and wing moult, July 87

Swallows collecting mud for nest building from a muddy pool by the stream in my garden, sketched from my studio window, Dron 22nd May 1987

111
House Martin and Sand Martins (juveniles)

House Martins and a juvenile Swallow
sunbathing on the slates of a neighbour's house,
Dron, 27th August 87

Moncrieffe Island, Perth 16/17th January 1987,
imm Cormorant & ♂ Goldeneye swimming in strong currents
with upwellings & whirlpools.

Whooper Swans, 26 adult and 5 young
eding in a snow covered winter barley field
Rhynd, near Perth, 16th January 87
mole hill

Whooper Swan family, Glen (river) Lyon, 30th January 1987

© Keith Brockie

Estuary

The mouth of the Tay estuary is bordered by the large sand dunes of Buddon to the north and Tentsmuir to the south. Low tide exposes a large area of sandbanks off Tentsmuir called the Abertay Sands, which are a favourite haul-out for many grey and common seals. Further in towards Tayport are extensive mussel banks and a shingle scar called Lucky Scalp (or Lucky Beacon – the stone beacon was removed some years ago because it was unsafe). Inland from the bridges the estuary becomes much more muddy. At low tide these mud banks are a favourite roost for thousands of pink-footed geese during the winter months. The northern shore nearly up to Perth and Mugdrum Island off Newburght is lined with dense Phragmites (very tall reeds) beds.

The estuary is of international importance for its large wintering flock of eider duck. This flock usually contains over 10,000 ducks, and, at its peak in November, up to 20,000 have been counted – some 20 per cent of the British population. Ringing recoveries have shown that they are mainly ducks breeding on the east coast of Scotland. To see this flock in flight, especially the boldly patterned black and white drakes, is an amazing spectacle. Sometimes one can pick out a rarity such as the king eider. I was fortunate to be on Lucky Scalp when a king eider floated by close in amongst its more numerous cousins. It was an adult male in eclipse plumage, just starting to moult back into his more colourful breeding plumage. King eiders are more usually found in Arctic waters – I have painted them a few times previously in Varangerfjord, North Norway – but a few are recorded elsewhere each year, mostly in Scottish waters.

Lucky Scalp is my favourite sketching place on the estuary. This shingle scar forms a small island at high tide, its size depending on the tidal high water mark. I have to be on the island at least three hours before high tide because a surrounding channel soon fills up with water. Likewise I have to wait a similar period after high tide to let the sea recede enough for me to get off. I set up my portable canvas hide above the high water mark and then waited for the birds to come in to roost as their feeding grounds gradually became covered by the incoming tide. Depending on the season large numbers of waders, gulls, ducks and cormorants roost here, sometimes within a few feet of the hide. I wrote the following after spending a few hours sketching on a lovely March Day:

> It's amazing how sounds carry across water to me, encapsulated in my canvas hide, on Lucky Scalp. I can hear the constant hubbub of human

View from the hide on Lucky Scalp, roosting cormorants, gulls, oystercatchers and curlews with Tentsmuir forest in the background.

interference. The staccato rattle of machine-gun fire on the Barry Buddon army firing range in competition with a noisy scrambler motorcycle on the nearer Fife shoreline. The clattering of a helicopter and jet noise from R A F Leuchars, the rumble of distant trains and the throbbing engines of the passing pilot boat. All carry over the calm waters that resound with the more sympathetic noise of hundreds of displaying eider drakes. However, I only subconsciously register these interruptions as I concentrate on the hundreds of birds roosting in front of my hide. Seaducks such as eider, common scoter, long-tailed duck and goldeneye come ashore to rest with their freshwater cousins the mallard, wigeon, teal and red-breasted merganser. Statuesque cormorants preen and dry themselves out in the weak sunshine. Dense flocks of curlew and oystercatcher sleep peacefully except when alerted by the occasional heron flying over, or alarmed by the sudden barking of dogs gambolling on the nearby beach. Offshore I can see two red-throated divers and huge flotillas of eider ducks. A gaudy male shellduck comes up to roost with his mate, his legs sporting five colour rings. A hidden voyeur amongst this gathering, I feel a deep contentment whilst engrossed in sketching.

I later learned that the sheldrake had been ringed by Durham University at Teesmouth on 6 December 1978. Subsequently it was seen at its moulting grounds on the Grosser Knechtsand in North Germany on 29 August 1981. Since then it hadn't been sighted till my observation at Lucky Scalp on 12 March 1987. This shows the value of colour marking individual birds to find out more about their movements. Another colour-ringed bird seen here was a cormorant aged over nineteen years old. Cormorants are wonderful birds to draw, immensely sculptural and full of character. I had great fun watching and sketching the immature bird attempting to swallow a large flatfish that bulged in its throat and neck. Its gape and neck must be very elastic in order to consume such an awkard fish. They look equally ridiculous drying out their plumage, especially when very wet.

Beyond Lucky Scalp lie the forest and sand dunes of Tentsmuir. The Nature Conservancy Council manages Tentsmuir Point as a National Nature Reserve for its mainly botanical interest. This is one of the few coastal areas which is increasing in size as a result of the steady accretion of sand and subsequent colonisation by vegetation. However, the Tentsmuir area is a mere vestige of its former state. Drainage and afforestation have largely destroyed the moorland and marshland habitat. Prior to the 1940s, when part of the area began to be used as an artillery range, large numbers of terns used to breed here. Once, as many as 2,000 pairs of common tern and 1,000 pairs of sandwich tern were counted. No terns now breed successfully as there is far too much human disturbance during the summer months. Up to 4,000 pairs of black-headed gull used to breed here, too, but they have disappeared because the wet areas that they favour have been progressively drained. The heather moorland also used to support dunlin and golden plover but the last nesting attempt of the latter was recorded in 1936. Red grouse were introduced in 1876 for sporting purposes, but the last recorded observation was as long ago as 1947.

Tentsmuir Point is a high tide roost for many gulls and waders as well as for terns in the late summer and autumn. With NCC permission I set up my hide amongst the sand and marram grass at the Point, planning to sketch some of the waders coming back from their Siberian breeding grounds still in their summer plumage. Up to 142 grey plover were present here in late August and early September, most of which were just starting to moult into their drab grey winter plumage. I succeeded in sketching details of two birds including one male still resplendent in his full summer dress. Earlier I had drawn a group of three immaculate male bar-tailed godwits still in their reddish-brown summer plumage. (This brought back fond memories of 1986 in Varangerfjord, North Norway, when I had spent many hours painting this species at its nest on the tundra.) The roosting godwits had started to moult out their wing coverts but otherwise were in perfect plumage. Other waders present at the roost included black-tailed godwit, whimbrel, ruff, knot, sanderling, dunlin, spotted redshank, redshank and oystercatcher.

On 7 September, whilst drawing the grey plover, I heard a commotion amongst the roosting birds. Looking out of the side window of the hide I was lucky to see a juvenile hobby speed over only a few feet above. This dashing falcon is a rare vagrant to Scotland and was probably on its way south from Scandinavia. A fierce west wind was blowing that day and I had difficulty keeping my telescope steady in the flapping hide. The roosting birds were braced against the wind. One greater black-backed gull which I sketched was sitting on an exposed point being blasted by fine drifting sand. At this time of year also, large flocks of sandwich terns are on the move: most of these terns have bred further south on the Farne Islands off Northumberland, and many of the adults are plagued by the incessant begging calls of accompanying offspring.

The extensive sandbanks off Tentsmuir are host to many grey and common seals. At low tide they haul themselves out on the sand to rest. The grey seals' nearest breeding colony is on the Isle of May where they pup from October to December. Outside this season they can be found in large numbers on the Abertay Sands. They generally keep much further out than the more inquisitive common seals. Dogs in particular seem to attract the latter's attention. The seals often follow walkers and their dogs along the shoreline keeping abreast only a few metres out in the water. Sometimes they 'porpoise' through the water with a lovely fluid action, at times leaping well clear. Some 500-plus common seals live in the Tay estuary and surrounds. Individuals have been recorded upriver as far as Stormontfield, a few miles north of Perth. The pups are born at the end of June and beginning of July. Unlike grey seals they are born with their adult type coat and can swim with the mother almost immediately. I was able to stalk up on some weaned pups out on the sandbanks. They had fallen asleep earlier and, as the tide had receded, were some distance from the water's edge. One of them had a shock when it woke up and found me staring at it eyeball to eyeball! It didn't take too kindly to me but I managed to sketch a portrait in between its angry charges.

Tentsmuir Point National Nature Reserve has a rich plant life, and efforts are being made to control the spread of encroaching birch scrub in order to maintain the status quo. I painted quite a few orchids here and, amongst the bordering forest, such plants as lesser twayblade and creeping lady's tresses. Some, like the coralroot orchid, have a very localized distribution in Scotland. The small specimens I painted were growing in a clearing amidst the pines in the dune slacks. Others growing in the damp birchwood were double the height, with a light green stem. Another strange plant growing in the forest at the edge of the dunes is the yellow bird's nest, which only grows in a couple of Scottish localities. It is saprophytic – that is, it feeds on decaying organic matter. Sprouting out of a carpet of pine needles in deep shade, its anaemic yellow-coloured spikes are topped with nodding, undistinctive flowers. The moonwort and adderstongue are two rather atypical ferns growing on the dune slacks, the latter in a wetter area. I also encountered the moonwort high up on the slopes of Ben Lawers amongst the more alpine flora. My favourite plant on Tenstmuir is the grass of Parnassus, a beautiful flower with five clearly veined petals. In some parts of the dune slacks they carpet the area together with the pink rosettes of common and seaside centaury.

On fine days at the appropriate time of year the area can be alive with butterflies. The two I have painted are the most obvious species, the grayling and dark green fritillary. Moths are also very common, especially the diurnal six-spot burnet moth. Look closely at the coralroot orchid painting and you will see its caterpillar in a cocoon attached to the stem.

Grayling, Hipparchia semele on Wild Thyme, Thymus serpyllum, 1/1. Tentsmuir Point NNR, 13th August 87. This is a common butterfly in this area.

Dark Green Fritillary, Mesoacidalia agloja 1/1
feeding on Rosebay Willowherb, Tentsmuir Point N.N.R
30th June 87

Common Centaury, Centaurium erythraea
with a pair of mating Six-spot Burnet Moths 1/1,
Tentsmuir dunes, 15th August 1987

Grass-of-Parnassus, Parnassia palustris 1/1
Tentsmuir Point NNR, 18th August 87, common here
behind the dune system

Coralroot Orchid, Corallorhiza trifida 1/1
Tentsmuir, Fife, 18th June 87
The right hand spike has the caterpillar of a six spot burnet moth in its pupation cocoon attached to it.

Creeping Lady's Tresses, Goodyera repens 1/1
Tentsmuir Forest, 28th July 1987, this orchid is
widespread in this area.

Lesser Twayblade, Listera cordata, 1/1
Tentsmuir, Fife 24th June 87
growing amongst a carpet of pine needles in the forest edge just behind the dunes. Mosquitos were almost unbearable !!

Common Twayblade, Listera ovata 1/1
Morton Lochs, Tentsmuir, Fife 30th June 87
Just starting to flower

Yellow Birdsnest, Monotropa hypopitys 1/1
Tentsmuir 25th July 87

Moonwort, Botrychium lunaria 1/1
4.7.87
on dune edge at Tentsmuir Point NNR
Adderstongue, Ophioglossum vulgatum,
4 & 5th July 1987
Tentsmuir Point NNR

In Scotland the brent goose occurs in small numbers during the winter months, especially the pale-bellied race. I sketched these confiding geese from my vehicle overlooking Carnoustie beach. My main ornithological interest was in the intermediate plumage of the sole immature bird. Artistically, the strong back lighting helped give some form to the velvety black neck.

brown
buff grey
grey

adult.

Carnoustie, 11th/12th February 87

drawings from 8 (7ad, 1imm) Pale-bellied Brent Geese, B.b.hrota

1st winter bird feeding on Enteromorpha (green seaweed)

diagnostic pale tips to wing coverts and distinct demarkation on belly between old juvenile plumage and new flank feathers

shaking & ruffling feathers.

scratching head

bathing Curlews, Lucky Beacon, Tayport.
18th September 1986

Oystercatchers at roost, Lucky Scalp
18th Sept 86

dozing on the water

♂ Common Scoter, Lucky Scalp, Tayport
27th January 1987

seldom seen ashore in a healthy condition, these birds came ashore to rest for a while, other diving ducks ashore included Long-tailed Duck and Goldeneye

yellow/orange orbital ring
brown iris

probably a 2nd yr ♂ due to brownish wing & dull body plumage

Heron, roosting on Lucky Scalp at high tide,
Tayport, 12th March 1987

♂ King Eider, an adult
still in eclipse plumage, sketched off
Lucky Scalp, Tayport, Fife, 6th October 1987.
Probably the same individual which summers on the Ythan in Aberdeenshire.
bill & 'shield' dull compared to a ♂ in full plumage.
no white thigh patch
as yet

♀ & eclipse ♂ Eider Ducks on the tideline, sketched Lucky Scalp, Tayport, 6th Oct 87

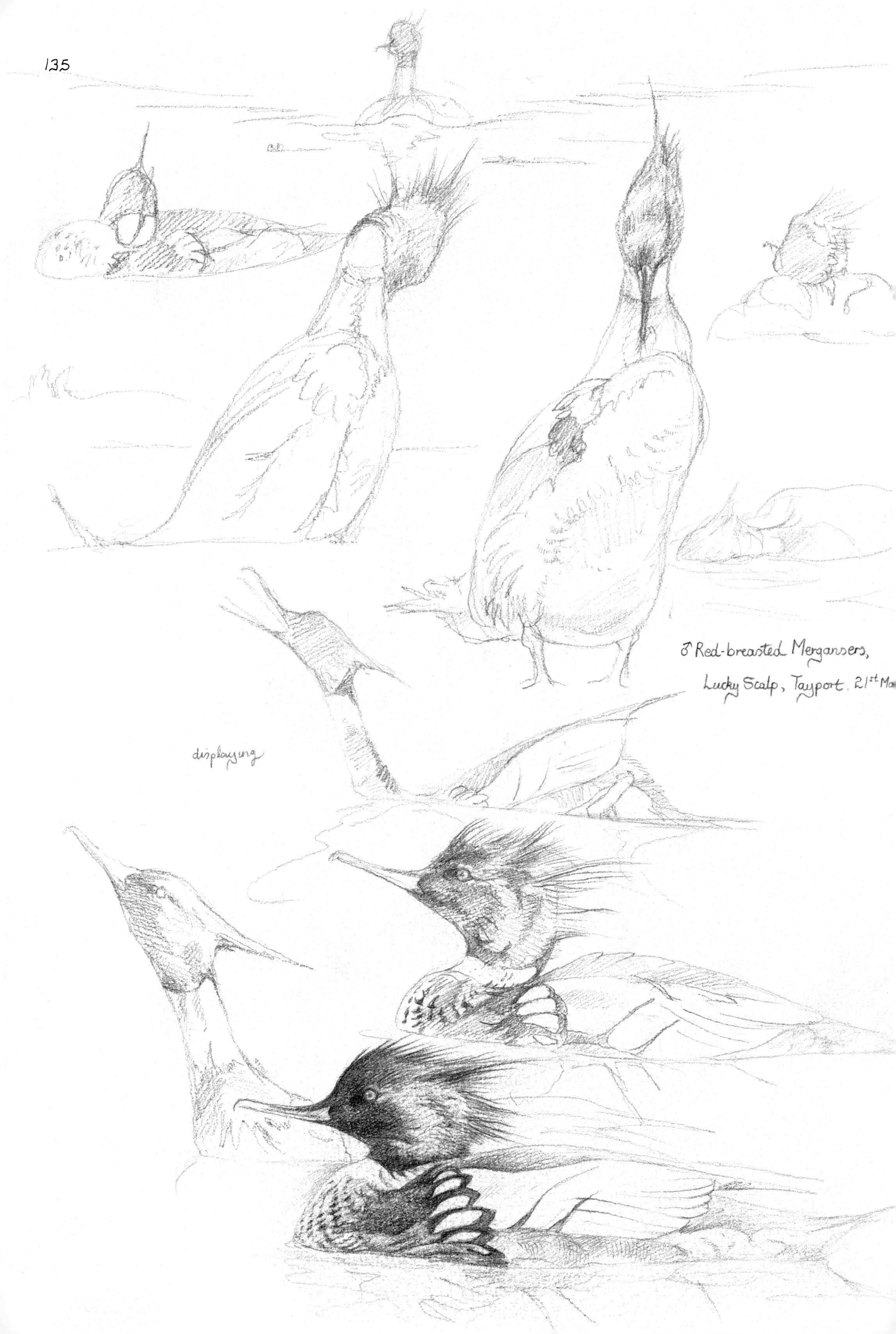
♂ Red-breasted Mergansers,
Lucky Scalp, Tayport. 21st Ma
displaying

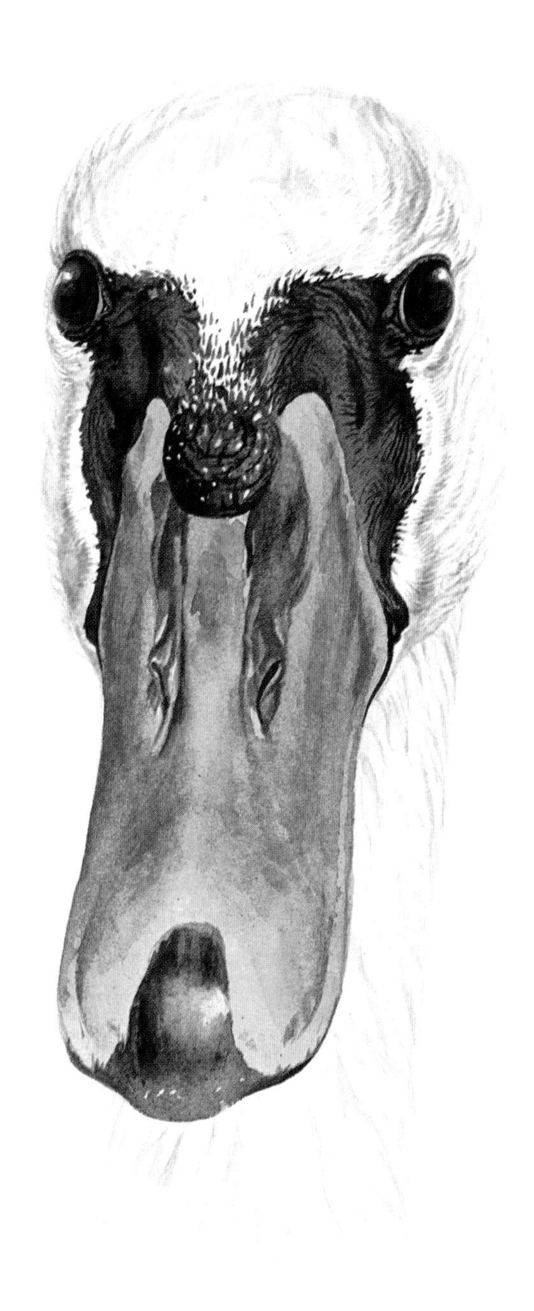

Mute Swan portrait, 1/1 8th September 87
Invergowrie Bay

imm
adults
Cormorants at roost;
dull flat light
Lucky Scalp, Tayport
27th January 1987

Immature Cormorant, Lucky Scalp, Tayport 27.1.87
bedraggled and fluffed out, drying itself

sketched from a temporary hide at the high tide roost

Adult Cormorant, Lucky Scalp, Tayport, 6th Oct 87
winter plumage, one of the many birds roosting less than
10 m's from my hide at high tide. This bird had a blue
colourring on and eventually I was able to read the number
on the metal ring – no M5579.

This bird had been ringed as a chick at Mochrum Loch,
Dumfries and Galloway on the 18th June 1968, so this
bird was 19 years 5 months old!

flatfish tail
distended throat
head
neck bulge
Juvenile Cormorant with its throat and neck bulging
with a flatfish which it couldn't swallow! It waddled up in
front of my hide and took a whole hour after many attempts
to swallow the fish. It was quite a sight whilst drying out
its wings topped by such a mishapen neck & head, it seemed
very relieved when it eventually swallowed the catch.
Lucky Scalp — 6th October 87

Adult and one second year Little Gulls,
Westhaven, Carnoustie 28th July 87
There were up to 400 present this week at this location in the early morning - oddly there were no juveniles at all.

Juvenile Arctic Tern, Tentsmuir, 3rd August 87,
sitting on the remains of an old salmon stake-net pole.

3 Bar-tailed Godwits, Tentsmuir Point, 25.8.87
(wing coverts in heavy moult)

© Keith Brockie

Grey Plovers moulting out of their summer plumage, Tentsmuir Point, 7th Sept 87

Sandwich Terns at roost, Tentsmuir

© Keith Brockie
August 87

Greater Black-backed Gull roosting on a sand-bar at
Tentsmuir Point being blasted by sand in the strong wind,
7th September 1987

Roosting Ringed Plover on the shore of the Tay estuary by Monifieth, 7th Nov 87, kelp in background.

Common Seals resting on a sandbank,
Tentsmuir, August 87

4 to 5 week old Common Seal pup
Tentsmuir, 29th July 87

I stalked up on this pup whilst it was sleeping on a sandbank, it had been left high and dry as the tide had receded. The fur on its upper head and back was dry contrasting markedly with the rest of the wet body.